Geographic Information Systems in Geospatial Intelligence

Edited by Rustam B. Rustamov

Published in London, United Kingdom

IntechOpen

Supporting open minds since 2005

Geographic Information Systems in Geospatial Intelligence
http://dx.doi.org/10.5772/intechopen.84925
Edited by Rustam B. Rustamov

Contributors
Dada Nade, Swapnil Potdar, Rani Pawar, Arnaud Le Bris, Nesrine Chehata, Debasish Chakraborty, Ali Dhafer Abed Yaha, Yashon Ombado Ouma, Andon Dimitrov Lazarov, Dimitar Minchev, Chavdar Minchev, Menachem Domb, Rustam B. Rustamov, Nasim Tohidi, Xavier Briottet, Nicolas Paparoditis

First published in London, United Kingdom, 2020 by IntechOpen
IntechOpen is the global imprint of INTECHOPEN LIMITED, registered in England and Wales, registration number: 11086078, 5 Princes Gate Court, London, SW7 2QJ, United Kingdom
Printed in Croatia

British Library Cataloguing-in-Publication Data
A catalogue record for this book is available from the British Library

Additional hard and PDF copies can be obtained from orders@intechopen.com

Geographic Information Systems in Geospatial Intelligence
Edited by Rustam B. Rustamov
p. cm.
Print ISBN 978-1-83880-504-3
Online ISBN 978-1-83880-505-0
eBook (PDF) ISBN 978-1-83969-129-4

For EU product safety concerns:
IN TECH d.o.o., Prolaz Marije Krucifikse Kozulić 3, 51000 Rijeka, Croatia, info@intechopen.com or visit our website at intechopen.com.

Meet the editor

Rustam B. Rustamov, PhD, specializes in space instrumentation and remote sensing and GIS technology. He obtained a PhD from the Russian Physical-Technical Institute (St. Petersburg). Dr. Rustamov was invited to work at the European Space Agency within the framework of the United Nations Programme on Space Applications at the European Space Research and Technology Centre, the Netherlands. He is appointed to the United Nations Office for Outer Space Affairs Action Teams, United Nations Economical and Social Commission for Asia and the Pacific, and the International Astronautical Federation, for which he currently serves as co-chair of the International Astronautical Congress. He is an author of sixteen books and more than 130 scientific papers.

Contents

Preface

There are obvious stages of satellite data collection and processing. In general, there are two modes of interaction between remote sensing and geographical information systems (GIS). Remote sensing can be used to generate digital maps that can be integrated into GIS development, whereas GIS data can be applied to interpret and classify remotely sensed data. There is no doubt that it is very important to find out reliable digital sources and point out the proper method for achieving high-accuracy data processing.

Remote sensing and GIS technology are used to improve satellite image processing and classification. Research in this area is linked to numerous factors that affect Earth monitoring such as natural resources, natural disaster observation, urban extension, and intensification of land use and land cover including deforestation, afforestation, land abandonment, and so on. As such, GIS and remote sensing represent useful tools for assessing/evaluating the detection of changes.

In recent years, however, more sophisticated data-driven methods have been used for Earth monitoring because they are more robust and have better capability to handle complicated relationships between input variables. It takes a vital place in use of current technology applications of different machine learning algorithms, including artificial neural networks (ANN), adaptive neuro-fuzzy inference systems (ANFIS), decision trees (DT), or support vector machines (SVM).

From this point of view, achievements in GIS applications are becoming widely important.

In chapter 1 SAR modeling of geophysics measurements is described for analyzing and modeling SAR interferometric processes in scenarios with different geometric, kinematics, and geological structures as well as for generating pseudo SAR interferograms based on geophysical measurements and topographic maps.

Chapter 2 of this book introduces various navigation implementations using alternate technologies integrated with GPS or operated as standalone devices for expanding navigation systems through combining advanced GIS data processing technologies.

Chapter 3 analyzes machine learning in GIS to develop the megacities application.

In chapter 4, we present research results related to the factors that affect high-accuracy data processing. To begin, we include a study of equatorial plasma bubbles using sky and GPS systems to measure total electron content (TEC) using a GPS receiver and images of the nightglow OI 630.0 nm emissions.

Chapter 5 describes the study of the spectral optimization of an airborne multispectral camera for land cover classification focuses on the choice of such relevance score. Several criteria are compared through both quantitative and qualitative analyses. To achieve a fair comparison, all tested criteria are compared to classic hyperspectral data sets using the same optimization heuristics: an incremental one to assess the impact of the number of selected bands and a stochastic one to obtain several possible good band subsets and to derive band importance measures out of intermediate good band subsets.

Chapter 6 highlights the Hölder exponent and variance-based clustering method for classifying land use/land cover in high spatial resolution, remotely sensed images with clustering techniques.

In Chapter 7, an integrated database of road design elements is used for exporting all the design elements to the GIS program by creating an integrated road database. The achieved database has capability of spatial analysis and connectivity, integrating other parts of the road network in the city.

Chapter 8 presents the results of research using low-cost RGB-D sensors for autonomous pothole detection with spatial fuzzy c-means segmentation. Results demonstrate the advantage of complementary processing of low-cost multisensory data, through channeling data streams and linking data processing according to the merits of the individual sensors, for autonomous cost-effective assessment of road-surface conditions using remote sensing technology.

Rustam B. Rustamov
EILINK Research and Development Center of Khazar University,
Baku, Azerbaijan

Chapter 1

InSAR Modeling of Geophysics Measurements

Andon Lazarov, Dimitar Minchev and Chavdar Minchev

Abstract

In the present work, the geometry and basic parameters of interferometric synthetic aperture radar (InSAR) geophysics system are addressed. Equations of pixel height and displacement evaluation are derived. Synthetic aperture radar (SAR) signal model based on linear frequency modulation (LFM) waveform and image reconstruction procedure are suggested. The concept of pseudo InSAR measurements, interferogram, and differential interferogram generation is considered. Interferogram and differential interferogram are generated based on a surface model and InSAR measurements. Results of numerical experiments are provided.

Keywords: InSAR, geometry, signal modeling, SAR interferogram, SAR differential interferograms

1. Introduction

Synthetic aperture radar (SAR) is a coherent microwave imaging instrument capable to provide for data all weather, day and night, guaranteeing global coverage surveillance. SAR interferometry is based on processing two or more complex valued SAR images obtained from different SAR positions [1–4]. The InSAR is a system intends for geophysical measurements and evaluation of topography, slopes, surface deformations (volcanoes, earthquakes, ice fields), glacier studies, vegetation growth, etc. The estimation of topographic height with essential accuracy is performed by the interferometric distance difference measured based on two SAR echoes from the same surface. Changes in topography (displacement), precise to a fraction of a radar wavelength, can be evaluated by differential interferogram generated by three or more successive complex SAR images [5, 6]. Demonstration of time series InSAR processing in Beijing using a small stack of Gaofen-3 differential interferograms is discussed in [7].

A general overview of the InSAR principles and the recent development of the advanced multi-track InSAR combination methodologies, which allow to discriminate the 3-D components of deformation processes and to follow their temporal evolution, are presented in [8]. The combination of global navigation satellite system (GNSS) and InSAR for future Australian datums is discussed in [9].

A high-precision DEM extraction method based on InSAR data and quality assessment of InSAR DEMs is suggested in [10, 11]. InSAR digital surface model (DSM) and time series analysis based on C-band Sentinel-1 TOPS data are presented in [12, 13]. DEM registration, alignment, and evaluation for SAR interferometry, deformation monitoring by ground-based SAR interferometry (GB-InSAR), a field test in dam, and an improved approach to estimate large-gradient deformation using high-resolution TerraSAR-X data are discussed in [14–16]. *InSAR Time-Series Estimation of the Ionospheric Phase Delay: An Extension of the Split Range-Spectrum Technique* and InSAR data coherence estimation using 2D fast Fourier transform are performed in [17, 18].

In comparison with the results described in the aforementioned publications, the main goal of the present work is to suggest an analytical model of multi-pass InSAR geometry and derive analytical expressions of current distances between SAR's positions and individual pixels on the surface and to describe principal InSAR parameters: topographic height and topographic displacement from the position of InSAR modelling. The focus is on the two modelling approaches: first, by the definition of real scenario, geometry, and kinematics and SAR signal models and corresponding complex image reconstruction and interferogram and differential interferogram generation and, second, the process of pseudo SAR measurements and interferogram generation that is analytically described. Results of numerical experiments with real data are provided.

The rest of the chapter is organized as follows. In Section 2, 3D InSAR geometry and kinematics are analytically described. In Section 3 and Section 4, analytical expressions of InSAR relief measurements and relief displacement measurements are presented. In Section 5 and Section 6, SAR waveform, deterministic signal model, and image reconstruction algorithm are described. In Section 7, numerical results of InSAR modelling based on the geometry, kinematics, and signal models are provided. In Section 8 and Section 9, a pseudo InSAR modelling of geophysical measurements and numerical results are presented, respectively. Conclusion remarks are made in Section 10.

2. InSAR geometry and kinematics

Assume a three-pass SAR system viewing three-dimensional (3-D) surface presented by discrete resolution elements, pixels. Each pixel is defined by the third coordinate $z_{ij}\left(x_{ij}, y_{ij}\right)$ in 3-D coordinate system Oxyz. Let A, B, and C, be the SAR positions of imaging. Between every SAR position, $C_3^2 = 3$ InSAR baselines can be drawn.

The basic geometric SAR characteristic is the time-dependent distance vector from SAR to each pixel on the surface in the n-th SAR pass at the p-th moment defined by

$$R_{ij}^n(p) = R^n(p) - R_{ij} = \left[x_{ij}^n(p), y_{ij}^n(p), z_{ij}^n(p)\right]^T, \qquad (1)$$

where n = 1–3 is the number of SAR passes and $R^n(p) = R^{0n} + V.p.T_p$ is the distance vector in the n-th SAR pass at the p-th moment, R^{0n} is the initial distance vector in the n-th SAR pass, R_{ij} is the constant distance vector of the ij- th pixel on the surface, and $x_{ij}^n(p)$, $y_{ij}^n(p)$, and $z_{ij}^n(p)$ are the current coordinates of $R_{ij}^n(p)$ written by the expression.

$$x_{ij}^n(p) = x^n(p) - x_{ij}, y_{ij}^n(p) = y^n(p) - y_{ij}, z_{ij}^n(p) = z^n(p) - z_{ij} \qquad (2)$$

where $x_{ij} = i\Delta X$, $y_{ij} = j\Delta Y$, and $z_{ij} = z_{ij}\left(x_{ij}, y_{ij}\right)$ is the pixel's discrete coordinates and $x^n(p)$, $y^n(p)$, and $z^n(p)$ are the SAR current coordinates in the n-th pass, defined by the following equation.

$$x^n(p) = x_0^n - V_x p T_p,\ y^n(p) = y_0^n - V_y p T_p,\ z^n(p) = z_0^n - V_z p T_p, \tag{3}$$

where x_0^n, y_0^n, and z_0^n are the SAR initial coordinates in the n-th pass, measured at the initial moment; T_p is the time repetition period; p is the number of the emitted pulse; $V = \left[V_x, V_y, V_z\right]^T$ is the SAR vector velocity; $V_x = V\cos\alpha$, $V_y = V\cos\beta$, and $V_z = V\cos\delta$ are the components of vector velocity; $\cos\alpha$, $\cos\beta$, and $\cos\delta = \sqrt{1 - \cos^2\alpha - \cos^2\beta}$ are the guiding cosines; and V is the module of the vector velocity V. Modulus of the current distance vector $R_{ij}^n(p)$ is defined by

$$R_{ij}^n(p) = \left\{\left[x_{ij}^n(p)\right]^2 + \left[y_{ij}^n(p)\right]^2 + \left[z_{ij}^n(p)\right]^2\right\}^{\frac{1}{2}}. \tag{4}$$

Eq. (4) can be used to model a SAR signal from the ij-th pixel in the n-th SAR pass by calculation of the respective time delay and phase of the signal.

3. InSAR relief measurements

The distances to ij-th pixel from SAR in m-th and n-th pass ($m \neq n$) at the moment of imaging can be defined by the cosine's theorem, i.e.,

$$\left|R_{ij}^n\right| = \left\{\left|R_{ij}^m\right|^2 + B_{mn}^2 - 2B_{mn}\left|R_{ij}^m\right| \cos\left[\frac{\pi}{2} - \left[\theta^{m_{ij}} - \alpha_{mn}\right)\right]\right\}^{\frac{1}{2}}, \tag{5}$$

where B_{mn} is the modulus of the baseline vector, $\theta^{m_{ij}}$ is the look angle, and α_{mn} is a priory known tilt angle, the angle between the baseline vector and plane Oxy. The look angle $\theta^{m_{ij}}$ and height h^m of an ij-th pixel on the surface with respect to m-th SAR position in the moment of imaging can be written as

$$\theta^{m_{ij}} = \alpha_{mn} + \arcsin\frac{\left|R_{ij}^m\right|^2 + B_{mn}^2 - \left|R_{ij}^n\right|^2}{2B_{mn}\left|R_{ij}^m\right|}, \tag{6}$$

$$z_{ij} = h^m - \left|R_{ij}^m\right|\cos\theta^{m_{ij}}. \tag{7}$$

The distance difference, $\left|\Delta R_{ij}^{mn}\right| = \left|R_{ij}^n\right| - \left|R_{ij}^m\right|$, can be expressed by the interferometric phase difference $\left|\Delta R_{ij}^{mn}\right| = \frac{\lambda}{2\pi}\Delta\phi_{ij}^{mn}$. In case $\left|R_{ij}^m\right|$ can be measured, i.e., $\left|R_{ij}^n\right| = \left|R_{ij}^m\right| + \left|\Delta R_{ij}^{mn}\right|$, then

$$\theta^{m_{ij}} = \alpha_{mn} + \arcsin\left[\frac{B_{mn}}{2R_{ij}^m} - \frac{\lambda}{2\pi B_{mn}}\Delta\phi_{ij}^{mn}\left(1 + \frac{\lambda}{4\pi R_{ij}^m}\Delta\phi_{ij}^{mn}\right)\right], \tag{8}$$

$$z_{ij} = h^m - R_{ij}^m.\cos\left\{\alpha_{mn} + \arcsin\left[\frac{B_{mn}}{2R_{ij}^m} - \frac{\lambda}{2\pi B_{mn}}\Delta\phi_{ij}^{mn}.\left(1 + \frac{\lambda}{4\pi R_{ij}^m}\Delta\phi_{ij}^{mn}\right)\right]\right\}. \tag{9}$$

4. InSAR measurements of relief displacement

Consider a three-pass SAR interferometry (**Figure 1**). Let A and B be the two positions of imaging which can be defined by two passes of the same spaceborne SAR in different time (two pass interferometry). The third position C is defined by the third pass of the spaceborne SAR. The surface displacement, Δz_{ij}, due, for instance, to an earthquake could derive from two SAR interferograms built before and after the seismic impact. The temporal baseline, the time scale over which the displacement is measured, must follow the dynamics of the geophysical phenomenon. Short-time baseline is applied for monitoring fast surface changes. Long temporal baseline is used for monitoring slow geophysics phenomena (subsidence). The interferometry phase before event is derived from complex images acquired by A and B SAR positions in the moment of imaging, while the interferometry phase after event is derived from complex images acquired by A and C SAR positions in the moment of imaging. The distances R_{ij}^1, R_{ij}^2, R_{ij}^3, and R_{ij}^{d3} after standard manipulations are written as.

$$R_{\mathrm{ij}}^2 \simeq R_{\mathrm{ij}}^1 - B_1 \sin\left(\theta_{\mathrm{ij}} - \alpha_1\right) + \frac{B_1^2}{2R_{\mathrm{ij}}^1}, R_{\mathrm{ij}}^3 \simeq R_{\mathrm{ij}}^1 - B_2 \sin\left(\theta_{\mathrm{ij}} - \alpha_2\right) + \frac{B_2^2}{2R_{\mathrm{ij}}^1}, \tag{10}$$

$$R_{\mathrm{ij}}^{\mathrm{d3}} \simeq R_{\mathrm{ij}}^3 - \Delta z\left(\cos\theta_{\mathrm{ij}} + \frac{B_2}{R_{\mathrm{ij}}^1}\sin\alpha_2\right) + \frac{\left(\Delta z_{\mathrm{ij}}\right)^2}{2R_{\mathrm{ij}}^1},$$

where R_{ij}^1, R_{ij}^2, and R_{ij}^3 are the slant ranges from A, B, and C positions of SAR system to the observed pixel in the moment of imaging before the surface

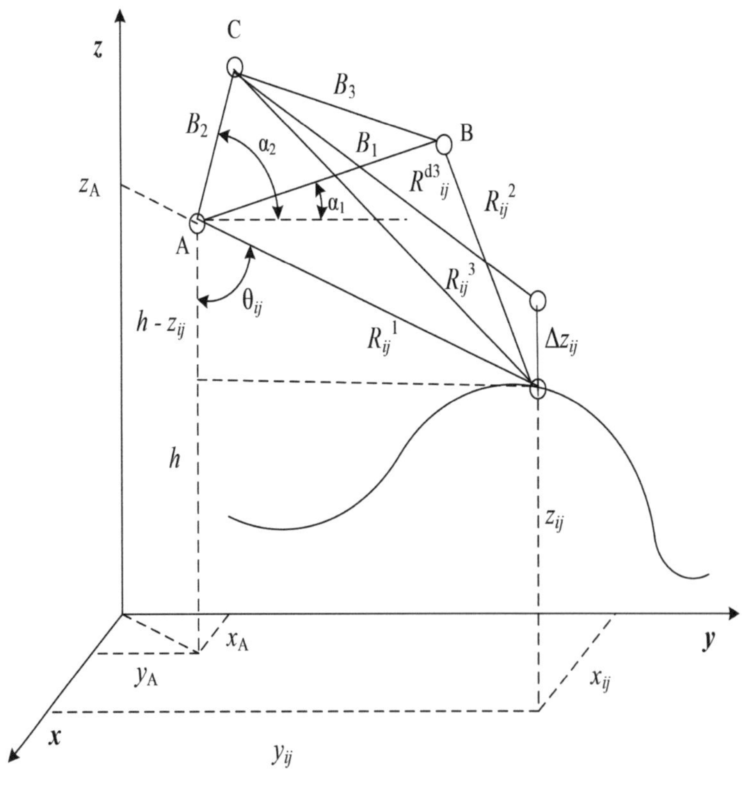

Figure 1.
InSAR geometry and kinematics.

displacement and $R_{\rm ij}^{d3}$ is the slant range to from C SAR position to the observed pixel after $\Delta z_{\rm ij}$ surface displacement.

Given the SAR wavelength λ, the phase differences proportional to range differences related to a particular pixel before and after displacement in the moment of imaging can be written as.

$$\phi^{\rm AB} = \frac{4\pi}{\lambda}\left(R_{\rm ij}^1 - R_{\rm ij}^2\right), \phi^{\rm AC} = \frac{4\pi}{\lambda}\left(R_{\rm ij}^1 - R_{\rm ij}^3\right), \phi_d^{\rm AC} = \frac{4\pi}{\lambda}\left(R_{\rm ij}^1 - R_{\rm ij}^{d3}\right) \quad (11)$$

Neglecting the term $(\Delta z)^2/2R_{\rm ij}^1$ in Eq. (10) can be rewritten as.

$$\phi^{\rm AB} = \frac{4\pi}{\lambda}\left(B_1 \sin\left(\theta_{\rm ij} - \alpha_1\right) - \frac{B_1^2}{2R_{\rm ij}^1}\right);\ \phi^{\rm AC} = \frac{4\pi}{\lambda}\left(B_2 \sin\left(\theta_{\rm ij} - \alpha_2\right) - \frac{B_2^2}{2R_{\rm ij}^1}\right) \quad (12)$$

$$\begin{aligned}\phi_d^{\rm AC} &= \frac{4\pi}{\lambda}\left[B_2 \sin\left(\theta_{\rm ij} - \alpha_2\right) - \frac{B_2^2}{2R_{\rm ij}^1} + \Delta z\left(\cos\theta_{\rm ij} + \frac{B_2}{R_{\rm ij}^1}\sin\alpha_2\right)\right] \\ &\phi^{\rm AC} + \Delta z\left(\cos\theta_{\rm ij} + \frac{B_2}{R_{\rm ij}^1}\sin\alpha_2\right)\end{aligned} \quad (13)$$

The displacement $\Delta z_{\rm ij}$ is extracted from the differential interferometric phase difference $\Delta\Phi_d = \phi_d^{\rm AC} - \phi^{\rm AB}$. Considering $B_2/R_{\rm ij}^1 < < 1$, then $\Delta\Phi_d = \Delta\Phi + \frac{4\pi}{\lambda}\Delta z_{\rm ij}\cos\theta_{\rm ij}$, where

$$\Delta\Phi = \frac{4\pi}{\lambda}\left[B_2 \sin\left(\theta_{\rm ij} - \alpha_2\right) - B_1 \sin\left(\theta_{\rm ij} - \alpha_1\right) - \frac{B_2^2 - B_1^2}{2R_{\rm ij}^1}\right]. \quad (14)$$

For surface displacement $z_{\rm ij}$ can be written as

$$\Delta z_{\rm ij} = \frac{\lambda}{4\pi}\frac{\Delta\Phi_d - \Delta\Phi}{\cos\theta_{\rm ij}}. \quad (15)$$

5. SAR waveform and deterministic signal model

The SAR transmits a series of electromagnetic waveforms to the surface, which are described analytically by the sequence of linear frequency modulation (chirp) pulses as follows

$$S(t) = \sum_{p=1}^{M} A \exp\left\{-j\left[\omega\left(t - {\rm pT}_p\right) + b\left(t - {\rm pT}_p\right)^2\right]\right\}, \quad (16)$$

where A is the amplitude of the transmitted pulses, T_p is the pulse repetition period, $\omega = 2\pi.c/\lambda$ is the angular frequency, $p = \overline{1, M}$ is the index of LFM emitted pulse, M is an emitted pulse number for synthesis of the aperture, $c = 3 \times 10^8$ m/s is the light speed in vacuum, ΔF is the LFM pulse bandwidth, $b = \pi.\Delta F/T$ is the chirp rate, and T is the time LFM pulse width.

The SAR signal, reflected by ij-th pixel and registered in the n-th pass, can be expressed as

$$S_{ij}^{n}(t) = a_{ij}(z_{ij})\,\text{rect}\frac{t - t_{ij}^{n}}{T}\,\exp\left\{-j\left[\omega\left(t - t_{ij}^{n}\right) + b\left(t - t_{ij}^{n}\right)^{2}\right]\right\} \tag{17}$$

$$\text{rect}\frac{t - t_{ij}^{n}(p)}{T} = \left\{1, 0 < \frac{t - t_{ij}^{n}(p)}{T} \leq 1\right|, \tag{18}$$

where $a_{ij}(z_{ij})$ is the reflection coefficient of the pixel from the surface.

The parameter $a_{ij}(z_{ij})$ is a function of surface geometry; $t_{ij}^{n}(p) = \frac{R_{ij}^{1}(p)+R_{ij}^{n}(p)}{c}$ is the time propagation of the reflected signal from the ij-th scattering pixel registered in the n-th pass.

SAR signal reflected from the entire illuminated surface is an interference of elementary signals of scattering pixels and can be written as

$$S^{n}(t) = \sum_{i}\sum_{j} a_{ij}(z_{ij})\,\text{rect}\frac{t - t_{ij}^{n}}{T}\,\exp\left\{-j\left[\omega\left(t - t_{ij}^{n}\right) + b\left(t - t_{ij}^{n}\right)^{2}\right]\right\}. \tag{19}$$

The time dwell t of the SAR signal return for each transmitted pulse p can be expressed as $t = t_{ijmin}^{n}(p) + k\Delta T$, where $k = \overline{k_{ijmin}^{n}(p), k_{ijmax}^{n}(p)}$ is the sample number of the SAR return measured on range direction in n-th pass, $k_{ijmin}^{n} = \text{int}\left[t_{ijmin}^{n}(p)/\Delta T\right]$, $k_{ijmax}^{n} = \text{int}\left[t_{ijmax}^{n}(p)/\Delta T\right]$, $\Delta T = 1/(2\Delta F)$ is the sample time width, and $k_{max}^{n}(p)$ is the number of the furthest range bin where SAR signal is registered in n-th pass. Hence, in discrete form SAR signal can be rewritten as

$$\dot{S}^{n}(k,p) = \sum_{i}\sum_{j} a_{ij}(z_{ij})\,\text{rect}\frac{t - t_{ij}^{n}}{T} \cdot \exp\left\{-j\left[\omega\left((k-1)\Delta T - t_{ij}^{n}(p)\right) + b\left((k-1)\Delta T - t_{ij}^{n}(p)\right)^{2}\right]\right\} \tag{20}$$

The expressions derived in Section 2 and Section 5 can be used for modeling the SAR signal return in case the satellites are moving rectilinearly in 3-D coordinate system.

6. SAR image reconstruction

The complex image reconstruction includes the following operations: frequency demodulation, range compression, coarse range alignment, precise phase correction, and azimuth compression. The frequency demodulation is performed by multiplication of Eq. (20) with a complex conjugated function $\exp\left\{j\left(\omega(k-1)\Delta T + b[(k-1)\Delta T]^{2}\right)\right\}$.

Thus, the range distributed frequency demodulated SAR return in n-th pass for p-th pulse can be written as

$$\hat{\dot{S}}^{n}(k,p) = \sum_{i}\sum_{j} a_{ij}(z_{ij})\,\text{rect}\frac{(k-1)\Delta T - t_{ij}^{n}}{T} \cdot \exp\left\{-j\left[\omega t_{ij}^{n}(p) + b\left((k-1)\Delta T - t_{ij}^{n}(p)\right)^{2}\right]\right\}. \tag{21}$$

The range compression of the LFM demodulated SAR signal is performed by cross correlation with a reference function, $\exp\left\{\text{j}b[(k-1)\Delta T]^{2}\right\}$

$$\dot{S}_R^n\left(\hat{k},p\right) = \sum_{k=1}^{K} \dot{\hat{S}}^n (k,p) \exp\left\{ \mathrm{jb}\left[\left(k-\hat{k}-1\right)\Delta T\right]^2 \right\} \tag{22}$$

where K is the full number of LFM samples, the range bins where SAR signal is registered, and by Fourier transform

$$\dot{S}_R^n\left(\hat{k},p\right) = \sum_{k=1}^{K} \dot{\hat{S}}^n (k,p). \exp\left(j\frac{2\pi k\hat{k}}{K_{\max}^n}\right), \tag{23}$$

for each $p = \overline{1,M}$ and $\hat{k} = \overline{1,K}$.

The range alignment and higher-order phase correction are beyond of the scope of the present work. The azimuth compression is accomplished by Fourier transform of the range compressed signal, $\dot{S}_R^n\left(\hat{k},p\right)$. The complex image extracted from the n-th pass data can be expressed as

$$\dot{I}^n\left(\hat{k},\hat{p}\right) = \sum_{p=1}^{M} \dot{S}_R^n\left(p,\hat{k}\right) \exp\left(j\frac{2\pi p\hat{p}}{M}\right), \tag{24}$$

for each $\hat{p} = \overline{1,M}$, $\hat{k} = \overline{1,K}$.

The complex SAR image extracted from the n-th pass data preserves phases defined by distances from the satellite to each pixel at the moment of imaging. Based on pixel phases and image co-registration, a complex interferograms and differential interferograms can be created.

7. InSAR modeling: numerical results

The SAR signal model and imaging algorithm are illustrated by results of numerical experiments. Consider three pass satellite SAR system with position coordinates at the moment of imaging as follows.

$$x_0^1 = 0 \text{ m}; y_0^1 = 10.10^3 \text{ m}, z_0^1 = 100.10^3 \text{ m}, x_0^2 = 0 \text{ m}, \; y_0^2 = 10,1.10^3 \text{ m}, \\ z_0^2 = 100.10^3 \text{ m},$$

$$x_0^3 = 0 \text{ m}, \; y_0^3 = 10,2.10^3 \text{ m}, z_0^3 = 100.10^3 \text{ m}.$$

Coordinates of vector-velocity of the satellite are $v_x = 0$ m/s, $v_y = -600$ m/s, and $v_z = 0$ m/s. The surface observed by the SAR system is modeled by the following equation

$$z_{\mathrm{ij}} = 3\left(1-x_{\mathrm{ij}}\right)^2 \exp\left[-x_{\mathrm{ij}^2} - \left(y_{\mathrm{ij}}+1\right)^2\right] - 10\left(\frac{x_{\mathrm{ij}}}{5} - x_{\mathrm{ij}^3} - y_{\mathrm{ij}^5}\right)\exp\left(-x_{\mathrm{ij}^2} - y_{\mathrm{ij}^2}\right) - \\ -\frac{1}{3}\exp\left[-\left(x_{\mathrm{ij}}+1\right)^2 - y_{\mathrm{ij}^2}\right] \tag{25}$$

where $x_{\mathrm{ij}} = i\Delta X, y_{\mathrm{ij}} = j\Delta Y, i = \overline{1,I}, \; j = \overline{1,J}, I$ = 128 pixels; J = 128 pixels; ΔX; ΔY- the spatial resolution of the pixels.

Normalized amplitude of reflected signals from every pixel $a_{\mathrm{ij}} = 0.001$. The spatial resolution of the pixel are $\Delta X = \Delta Y = 2$ m. Wavelength is 0.03 m. Carrier frequency is 3.10^9 Hz. Frequency bandwidth is $\Delta F = 250$ MHz. Pulse repetition period is $T_p = 25.10^{-3}$ s. LFM pulse duration is $T = 5.10^{-6}$ s. Sample time duration is $\Delta T = 1,95.10^{-8}$ s. LFM sample number is K = 512. Emitted pulse number is

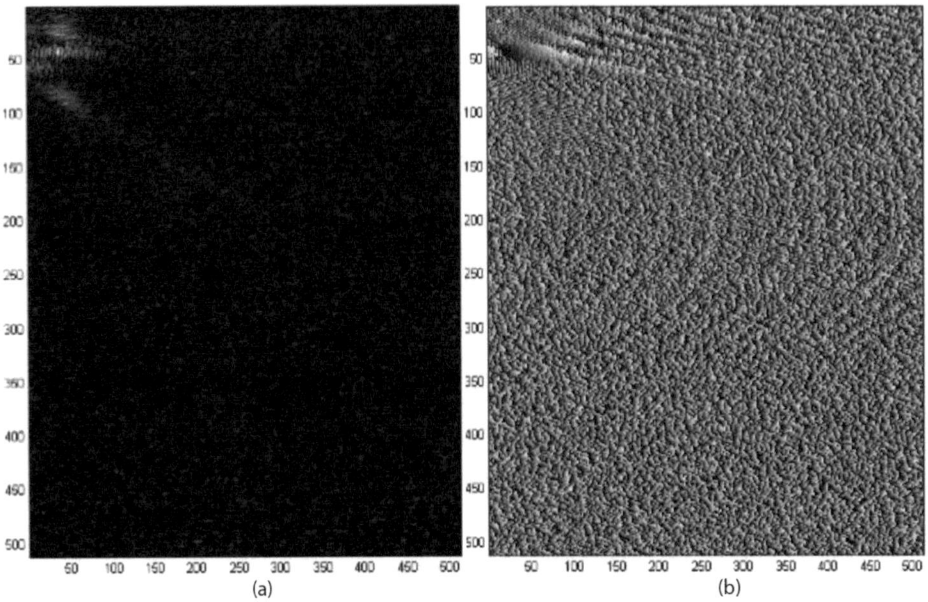

Figure 2.
The real (a) and imaginary (b) component of the SAR complex signal measured in the first SAR pass.

$M = 512$. Digital geometry description and SAR signal modeling are performed based on the theory in Sections 3 and 4. The complex images are extracted from the SAR signal by applying correlation range compression and FFT azimuth compression. Based on a priori-known kinematical parameters of satellites and coordinates of reference point from the surface autofocusing phase correction of the SAR signals registered in the both passes can be implemented.

The real and imaginary components of the SAR complex signal measured in the first SAR pass are depicted in **Figure 2**.

The complex SAR image's amplitude and phase obtained in the first SAR pass are depicted in **Figure 3**. The orientation of the surface's image (**Figure 3a**) in the frame is defined by the position of the SAR at the moment of imaging.

The real and imaginary components of the SAR complex signal measured in the second SAR pass are depicted in **Figure 4**.

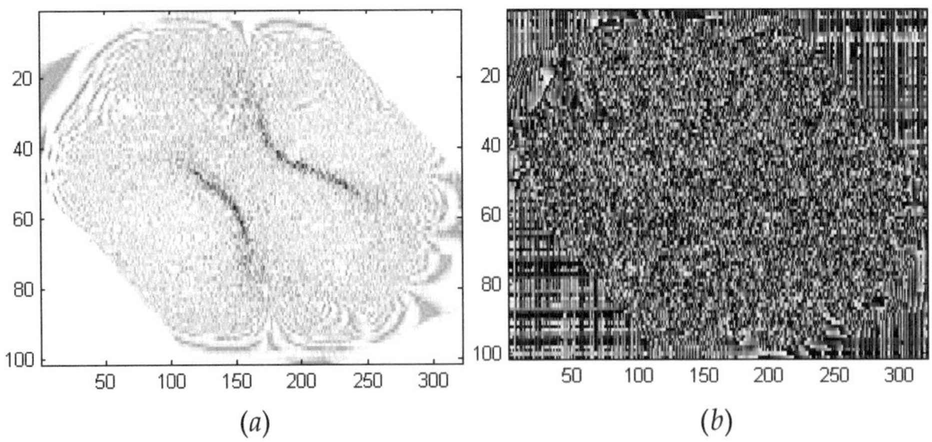

Figure 3.
The amplitude (a) and phase (b) component of the SAR complex image obtained in the first pass.

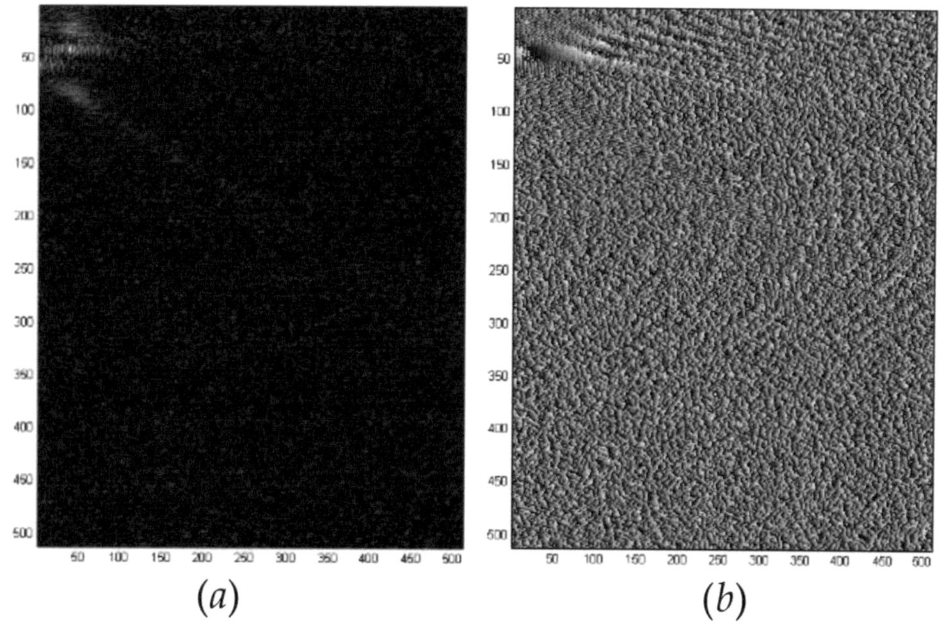

Figure 4.
The real (a) and imaginary (b) component of the SAR complex signal measured in the second SAR pass.

The complex SAR image's amplitude and phase obtained in the second SAR pass are depicted in **Figure 5**. It can be seen that the shape of the surface (the amplitude of the complex image) is similar to the shape of the surface obtained by the first SAR pass. In contrast, the phase structures of both complex images are different based on the different SAR positions in respect of the surface in the first and second pass at the moment of imaging.

By co-registration of the first and third SAR complex images, a complex SAR interferogram can be created with components in a coherent map and interferometric phase depicted in **Figure 6**.

The real and imaginary components of the SAR complex signal obtained in the third SAR pass is depicted in **Figure 7**.

The complex SAR image's amplitude and phase obtained in the third SAR pass are depicted in **Figure 8**. The shape of the surface obtained in the third SAR pass is

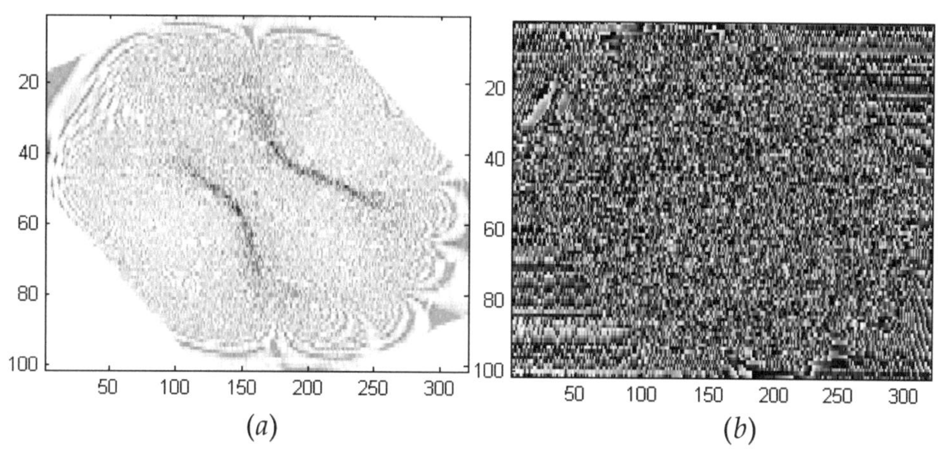

Figure 5.
The amplitude (a) and phase (b) components of the SAR complex image obtained in the second pass.

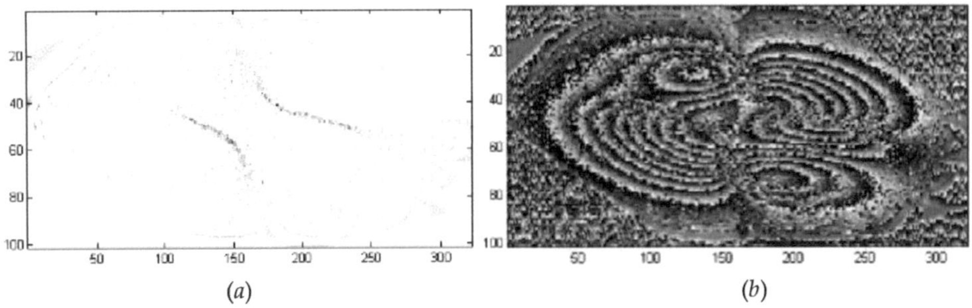

Figure 6.
The coherent map (a) and interferometric phase (b) of the complex SAR interferogram created by the first and second SAR complex images.

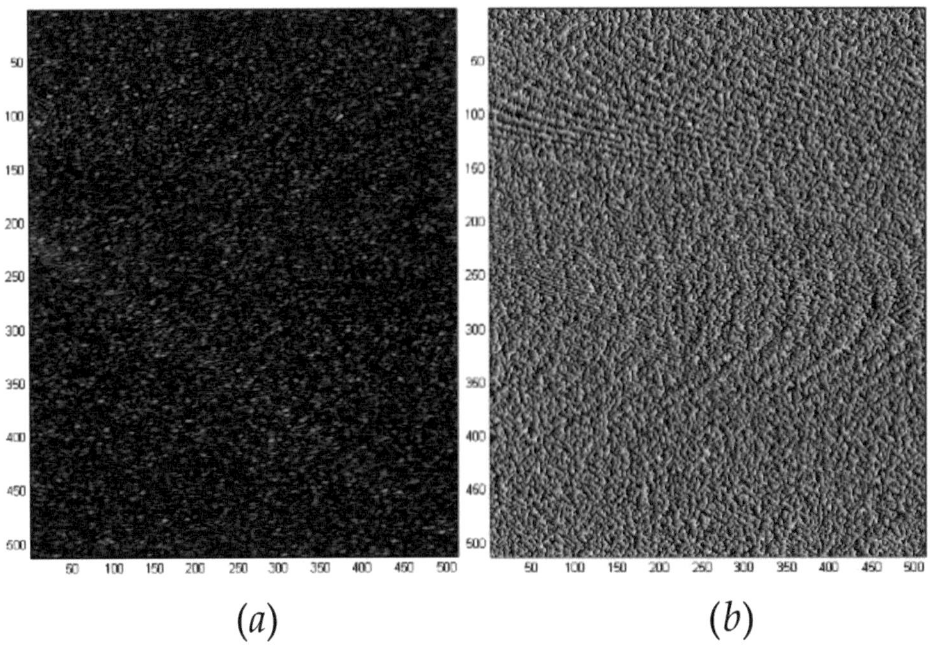

Figure 7.
The real (a) and imaginary (b) component of the SAR complex signal measured in the third SAR pass.

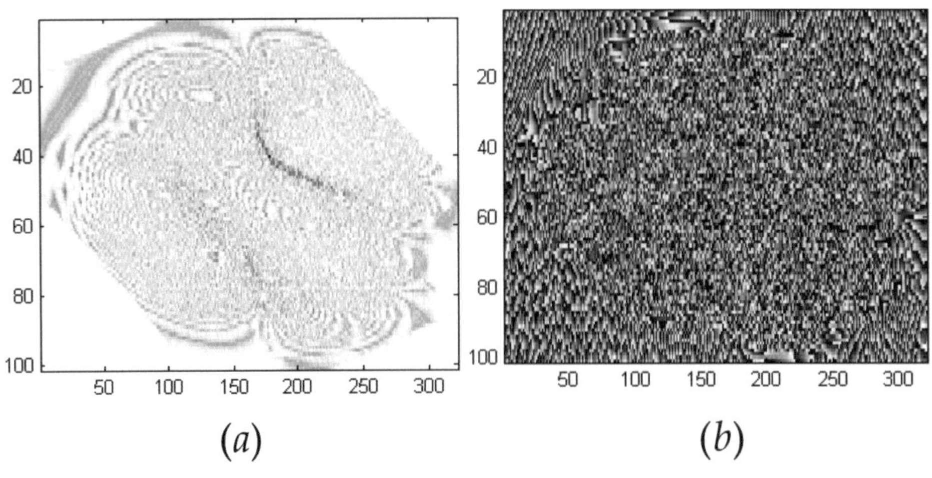

Figure 8.
The amplitude (a) and phase (b) component of the SAR complex image obtained in the third SAR pass.

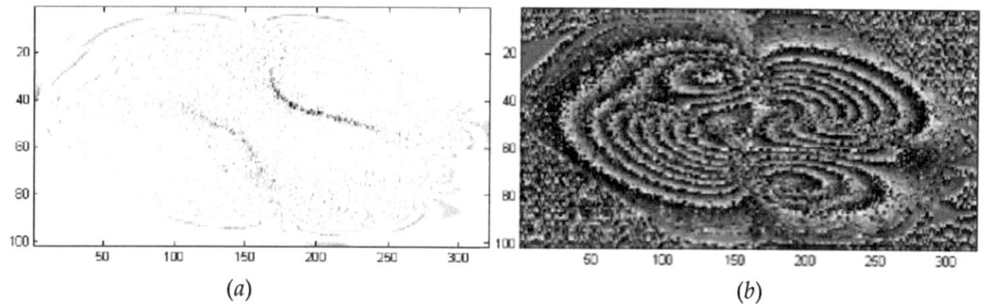

Figure 9.
The coherent map (a) and interferometric phase (b) of the complex SAR interferogram created by the first and third SAR complex images.

similar to the shape of the surface obtained by the first and second SAR passes. Comparing phase structures of the three complex SAR images, it can be noticed that they are different based on the different SAR's positions in respect to the surface at the moment of imaging.

Under pixel co-registration of the first and third SAR complex images, a complex SAR interferogram can be created with components in a coherent map and interferometric phase depicted in **Figure 9**.

Due to precise under pixel co-registrations of the first and second and the first and third SAR complex images, the phase interferograms depicted in **Figures 6b** and **9b**, respectively, are characterized with the similar structures.

8. Pseudo InSAR modeling of geophysical measurements

Consider three-pass InSAR geometry (**Figure 1**). The vector distances from the SAR positions to each ij-th pixel from the region of interest are $R_{\mathrm{ij}}^{S} = R^{S} - R_{\mathrm{ij}}$, where $S = A, B, C$ denotes the SAR position at the moment of imaging, $R^{S} = \left[x_S, y_S, z_S\right]^T$ denotes the SAR vector position, and $R_{\mathrm{ij}} = \left[x_{\mathrm{ij}}, y_{\mathrm{ij}}, z_{\mathrm{ij}}\right]^T$ denotes the ij-th pixel vector position. Coordinates of SAR positions in the moment of imaging are as follows: for a master SAR position A, x_A, y_A, z_A; for a slave SAR position B, x_B, y_B, z_B; and for a slave SAR position C, x_C, y_C, z_C.

After distance measurements from the master SAR position A and slave SAR positions B and C, respectively, to each ij-th pixel on the surface and co-registration of so obtained master image and slave images, the instrumental interferometric phase differences are calculated as follows

- without pixel displacement

$$\phi_{\mathrm{ij}}^{\mathrm{AB}} = \frac{4\pi}{\lambda}\left(\left|R_{\mathrm{ij}}^{A}\right| - \left|R_{\mathrm{ij}}^{B}\right|\right) - 2\pi.\max\left[\frac{2}{\lambda}\left(\left|R_{\mathrm{ij}}^{A}\right| - \left|R_{\mathrm{ij}}\right|\right)\right],$$

$$\phi_{\mathrm{ij}}^{\mathrm{AC}} = \frac{4\pi}{\lambda}\left(\left|R_{\mathrm{ij}}^{A}\right| - \left|R_{\mathrm{ij}}^{C}\right|\right) - 2\pi.\max\left[\frac{2}{\lambda}\left(\left|R_{\mathrm{ij}}^{A}\right| - \left|R_{\mathrm{ij}}^{C}\right|\right)\right]$$

- with pixel displacement

$$\phi_{\mathrm{ij},d}^{\mathrm{AC}} = \frac{4\pi}{\lambda}\left(\left|R_{\mathrm{ij}}^{A}\right| - \left|R_{\mathrm{ij},d}^{C}\right|\right) - 2\pi.\max\left[\frac{2}{\lambda}\left(\left|R_{\mathrm{ij}}^{A}\right| - \left|R_{\mathrm{ij},d}^{C}\right|\right)\right].$$

In order to unwrap the interferometric phases, standard algorithms, MATLAB *unwrap* function, 2-D Costantini phase unwrapping based on network programming, and 2D Goldstein branch cut phase unwrapping, can be applied.

9. Pseudo InSAR geophysical measurements: numerical results

Consider a GeoTIFF file of Dilijan region in Caucasus, Armenia, located at the geographical coordinates 40° 44′ 27″ north and 44° 51′ 47″ east longitude. Consider 2-pass InSAR scenario. Coordinates of SAR positions in the moment of imaging are the following: master SAR position A, $x_A = 0$ m, $y_A = 300.3 \times 10^3$ m, $z_A = 3 \times 10^5$ m and slave SAR position B $x_B = 0$ m, $y_B = 300 \times 10^3$ m, $z_B = 3 \times 10^5$ m. Wavelength is 0.05 m. Distances at the moment of imaging from the SAR position A and SAR position B to each pixel on the surface are illustrated in **Figure 10a, b**. Interferogram wrapped phases and unwrapped phases are presented in **Figure 10c, d**, respectively.

Consider a three-pass InSAR scenario and a surface before (**Figure 11a**) and after (**Figure 11b**) displacement described by MATLAB function *peaks*. Coordinates of SAR positions in the moment of imaging are the following: master SAR position A, $x_A = 350$ km, $y_A = 350$ km, $z_A = 800$ km; slave SAR position B $x_B = 351.5$ km,

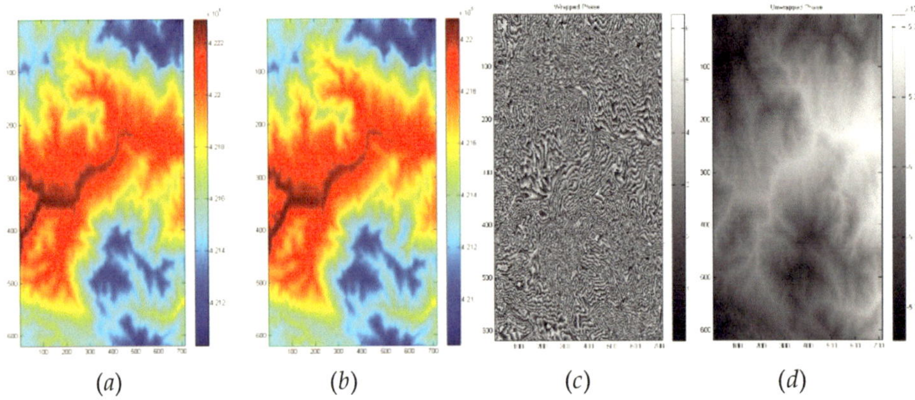

Figure 10.
Distances from SAR position A (a) and SAR position B (b) to each pixel on the surface in pseudo color map, interferogram wrapped phases (c) and unwrapped phases (d).

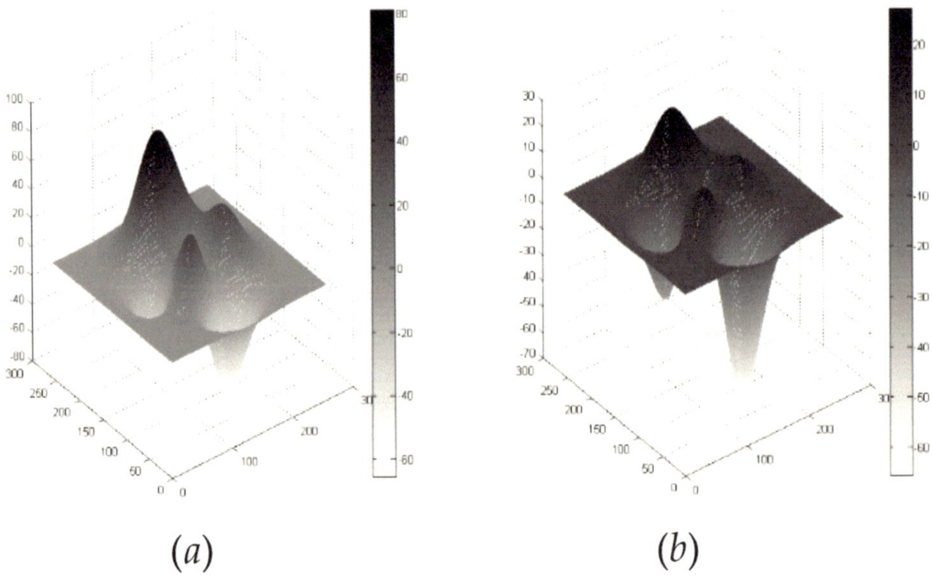

Figure 11.
Surface (peaks) before (a) and after (b) displacement.

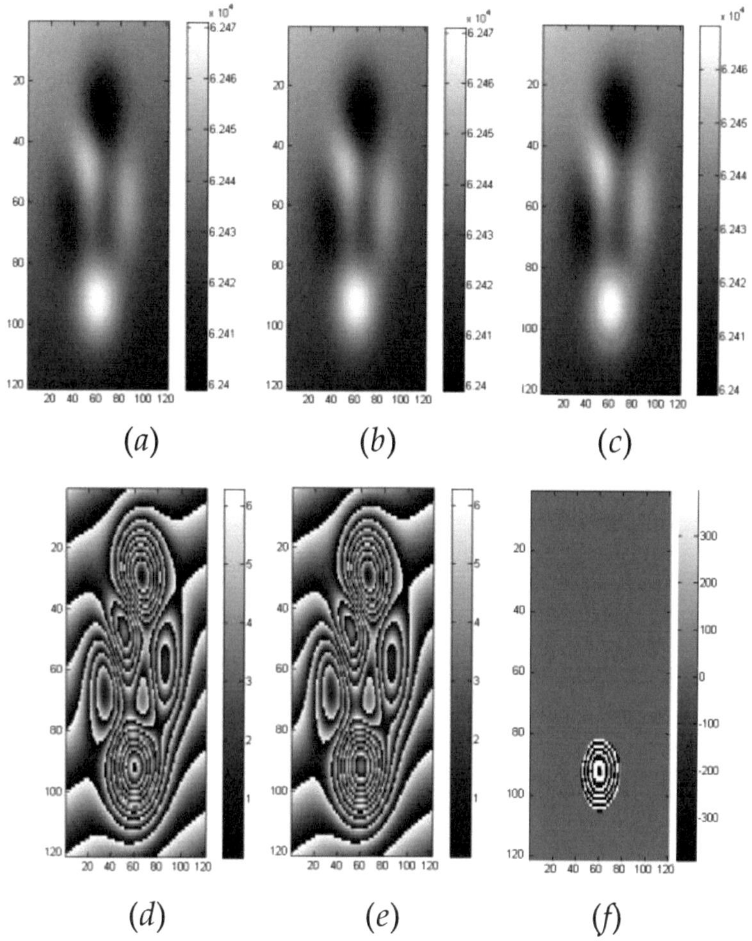

Figure 12.
Distances to the surface measured from SAR positions A (a), B (b), and C (c). AB interferogram (d), AC interferogram (e) with surface displacement, and differential interferogram AB-AC (f).

$y_B = 350$ km, $z_B = 800$ km; and slave SAR position C $x_C = 350$ km, $y_C = 351.2$ km, $z_C = 800$ m. Wavelength is 0.03 m.

Distances to the surface at the moment of imaging as pseudo collar maps measured from SAR positions A, B, and C are presented in **Figure 12a–c**, respectively. AB interferogram without surface displacement and AC interferogram with surface displacement are presented in **Figure 12d, e**, respectively. Differential interferogram AB-AC is presented in **Figure 12f**.

The differential interferogram obtained by pixel subtraction of interferograms in **Figure 12d, e** is presented in **Figure 12f**. It illustrates the displacement of the surface. Only deformed part of the surface as differential fringes is depicted. The pseudo InSAR modeling can be applied to generate interferograms and differential interferograms based on real geophysical measurements and Geo TIFF maps of the observed surface.

10. Conclusions

A multi-pass InSAR system has been theoretically analyzed and numerically experimented. Geometry and kinematics of multi-pass InSAR scenario have been analytically described. Mathematical expressions for definition of current distance vectors between SAR system and surface's pixels are derived. The basic InSAR parameters are defined. Analytical expressions to calculate pixel heights and pixel displacement have been derived. A model of linear frequency modulated SAR signal, reflected from the topographic surface, has been developed. An image reconstruction algorithm has been described. Numerical results verifying InSAR geometry, kinematics, and signal models are provided. Based on geometrical, kinematical, and signal models, numerical interferograms of a topographic surface have been created.

A pseudo InSAR approach has been applied to model processes of interferograms and differential interferogram generation using GeoTIFF files and measurements of distances from SAR positions to each pixels of the observed surface at the moment of imaging. Based on distance vector description of the InSAR scenario, the interferometric phase and interferometric differential phase have been analytically described. Pseudo InSAR geophysical measurements and interferograms and differential interferogram generation have been illustrated by results of numerical experiments.

In conclusion, the results in the present work can be applied for analysis and modeling of SAR interferometric processes in scenarios with different geometric, kinematics, and geological structures as well as for generating pseudo SAR interferograms based on the geophysical measurements and topographic maps.

Conflict of interest

The author declares no conflict of interest.

Author details

Andon Lazarov[1]*, Dimitar Minchev[2] and Chavdar Minchev[3]

1 Nikola Vaptsarov Naval Academy, Vasil Drumev, Varna, Bulgaria

2 Burgas Free University, Burgas, Bulgaria

3 AADCIS Faculty, Military University, Shumen, Bulgaria

*Address all correspondence to: lazarov@bfu.bg

References

[1] Pepe A, Bonano M, Zhao Q, Yang T, Wang H. The use of C-/X-band time-gapped SAR data and geotechnical models for the study of Shanghai's ocean-reclaimed lands through the SBAS-DInSAR technique. Remote Sensing. 2016;**8**:911. DOI: 10.3390/rs8110911

[2] Hu J, Ding X, Li Z, et al. Kalman-filter based approach for multi-sensor, multitrack and multitemporal InSAR. IEEE Transactions on Geoscience and Remote Sensing. 2013;**51**:4226-4239. DOI: 10.1109/TGRS.2012.2227759

[3] Casu F, Manconi A, Pepe A, Lanari R. Deformation time-series generation in areas characterized by large displacement dynamics: The SAR amplitude pixel-offset SBAS technique. IEEE Transactions on Geoscience and Remote Sensing. 2011;**49**:2752-2763. DOI: 10.1109/TGRS.2010.2104325

[4] Samsonov S, d'Oreye N. Multidimensional time-series analysis of ground deformation from multiple InSAR data sets applied to Virunga Volcanic Province. Geophysical Journal International. 2012;**191**:1095-1108. DOI: 10.1111/j.1365-246X.2012.05669.x

[5] Sansosti E, Berardino P, et al. How second-generation SAR systems are impacting the analysis of ground deformation. International Journal of Applied Earth Observation and Geoinformation. 2014;**28**:1-11. DOI: 10.1016/j.jag.2013.10.007. Available from: https://www.sciencedirect.com/science/article/pii/S0303243413001323 [Accessed: 11 August 2019]

[6] Vettore A, Ponte S, et al. Space-based surface change detection with differential Synthetic Aperture Radar (SAR) Interferometry: Potentialities and preliminary investigations. In: Symposium on Geospatial Theory, Processing and Applications; Ottawa; 2002. Available from: https://www.semanticscholar.org/paper/Space-based-surface-change-detection-with-Synthetic-Ponte-Crocetto/10591de3ab1c236a30f2eeaff3ab1f3e8da1af17 [Accessed: 11 August 2019]

[7] Wang J, Yu W, Deng Y, Wang R, et al. Demonstration of time-series InSAR processing in Beijing using a small stack of Gaofen-3 differential interferograms. Journal of Sensors. 2019;**2019**:4204580. DOI: 10.1155/2019/4204580

[8] Pepe A, Calò F. Review of interferometric synthetic aperture radar (InSAR) multi-track approaches for the retrieval of Earth's surface displacements. Applied Sciences. 2017;7:1264. DOI: 10.3390/app7121264

[9] Fuhrmann T, Garthwaite M, Lawrie S, Brown N. Combination of GNSS and InSAR for future Australian datums. In: International Global Navigation Satellite Systems Association IGNSS Symposium 2018; Colombo Theatres, Kensington Campus, UNSW Australia; 7–9 February 2018. Available from: http://www.ignss2018.unsw.edu.au/sites/ignss2018/files/u80/Papers/IGNSS2018_paper_17.pdf [Accessed: 11 August 2019]

[10] Xinshuang W, Lingling L, Xiaoliang Sh, Xitao H, Weil G. A high precision DEM extraction method based on InSAR data. In: ISPRS Annals of the Photogrammetry, Remote Sensing and Spatial Information Sciences, Volume IV-3, 2018 ISPRS TC III Mid-Term Symposium "Developments, Technologies and Applications in Remote Sensing", 7–10 May, Beijing, China; 2018. Available from: https://pdfs.semanticscholar.org/7f92/aca4273b2afb1b9347724d72eb4c13bcd1cc.pdf [Accessed: 10 August 2019]

[11] Crosetto M, Crippa B. Quality assessment of interferometric SAR DEMs. In: International Archives of

Photogrammetry and Remote Sensing. Vol. XXXIII, Part B1; Amsterdam; 2000. Available from: https://pdfs.semanticscholar.org/376f/4f7f12602827801b954c7fb79aae2585b9c3.pdf [Accessed: 28 July 2019]

[12] Letsios V, Faraslis I, Stathakis D. InSAR digital surface model (DSM) using Sentinel 1 and spatial data creation. In: AGILE 2019—Limassol; June 17–20, 2019. Available from: https://agile-online.org/images/conference_2019/documents/short_papers/50_Upload_your_PDF_file.pdf [Accessed: 28 July 2019]

[13] Liang C, Agram P, Simons M, Fielding EJ. Ionospheric correction of InSAR time series analysis of C-band Sentinel-1 TOPS data. This paper is a non-peer reviewed preprint published at EarthArXiv and submitted for consideration by IEEE TGRS. DOI: 10.1109/TGRS.2019.2908494. Available from: https://eartharxiv.org/atxr7/download

[14] Li Z, Bethel J. DEM registration, alignment and evaluation for SAR interferometry. In: The International Archives of the Photogrammetry, Remote Sensing and Spatial Information Sciences. Vol. XXXVII. Part B1. Beijing; 2008. pp. 11-116. Available from: https://www.isprs.org/proceedings/XXXVII/congress/1_pdf/19.pdf [Accessed: 07 August 2019]

[15] Mao W-J, Chang W-L. Deformation monitoring by ground-based SAR interferometry (GB-InSAR): A field test in dam. Advances in Information Sciences and Service Sciences (AISS). 2015;7(2):133-140. Available from: http://www.globalcis.org/aiss/ppl/AISS3808PPL.pdf [Accessed: 05 August 2019]

[16] Liu D, Chen R, Riedel B, Niemeier W. An improved approach to estimate large-gradient deformation using high resolution TerraSAR-X data. Solid Earth Discussions. 2014;**6**: 2759-2778. DOI: 10.5194/sed-6-2759-2014. Available from: https://www.solid-earth-discuss.net/6/2759/2014/sed-6-2759-2014.pdf [Accessed: 01 August 2019]

[17] Fattahi H, Simons M, Agram P. InSAR time-series estimation of the ionospheric phase delay: An Extension of the split Range-Spectrum technique. IEEE Transactions on Geoscience and Remote Sensing. 2017;**55**(2):777-786. Available from: http://web.gps.caltech.edu/~simons/publications/pdfs/Fattahi_2017.pdf [Accessed: 01 August 2019]

[18] Sosnovsky A, Kobernichenko V, Vinogradova N, Tsogtbaatar Ol. InSAR data coherence estimation using 2D fast Fourier transform. CEUR Workshop Proceedings (CEUR-WS.org). 2017; **1814**:98-105. Available from: http://ceur-ws.org/Vol-1814/paper-12.pdf [Accessed: 10 August 2019]

Chapter 2

Expanding Navigation Systems by Integrating It with Advanced Technologies

Menachem Domb

Abstract

Navigation systems provide the optimized route from one location to another. It is mainly assisted by external technologies such as Global Positioning System (GPS) and satellite-based radio navigation systems. GPS has many advantages such as high accuracy, available anywhere, reliable, and self-calibrated. However, GPS is limited to outdoor operations. The practice of combining different sources of data to improve the overall outcome is commonly used in various domains. GIS is already integrated with GPS to provide the visualization and realization aspects of a given location. Internet of things (IoT) is a growing domain, where embedded sensors are connected to the Internet and so IoT improves existing navigation systems and expands its capabilities. This chapter proposes a framework based on the integration of GPS, GIS, IoT, and mobile communications to provide a comprehensive and accurate navigation solution. In the next section, we outline the limitations of GPS, and then we describe the integration of GIS, smartphones, and GPS to enable its use in mobile applications. For the rest of this chapter, we introduce various navigation implementations using alternate technologies integrated with GPS or operated as standalone devices.

Keywords: navigation, GPS, IoT, GIS, public transport, smartphone application, UAVs, MAV, indoor/outdoor navigation, vision-based navigation, obstacle detection, SURF, control system, obstacle avoidance, pedestrian navigation system, modified RHKF filter, indoor assistive navigation, context-aware, semantic map, obstacle avoidance, tango device, near-field communication, indoor navigation, indoor positioning, mobile, NFC internal, sensor fusion, map matching, hidden Markov models, Kalman filter

1. The limitations of GPS

Some of the downsides of GPS are listed in [1]. Among these are several limitations which are relevant to this chapter. The weak intensity signal causes GPS to be less applicable for cases where stable navigating is mandatory or cases where navigating at indoor and covered areas. The low granularity of the signal accuracy makes navigation in crowded cities where landmarks are so close such that GPS is not able to differentiate among them and so is not effective. Furthermore, GPS signal may disperse and change its direction due to interruptions caused by skyscrapers, trees, geomagnetic storms, etc. The impact of unreliable GPS is huge especially due to the constant growing use of navigation applications such as Google Maps and Waze, which heavily rely on GPS signal. The impact may be more car accidents in cases where required information is missed exactly at the time it is critical and useful for driving continuation. GPS signal is not enough for covering all navigation instances. Local and timely knowledge is required for updated and accurate information to be able to properly react instantly when obstacles appear in the road ahead, for example, whether a deep pit or a flooding road is likely to be or if the road is closed. To reduce the dependency on GPS, several methods and technologies have been proposed, such as detailed map information, data from sensors, vision-based measurements, stop lines, and GPS-fused SLAM technologies.

2. GPS and GIS integration

A geographic information system (GIS) is a system designed to capture, store, manipulate, analyze, manage, and present all types of geographical data. Many electronic navigation systems deliver its road-guiding instructions using just verbal commands referring to the associated electronic map displayed to the user. This approach assumes the user familiarity with street maps and road networks, which sometimes is not so. In addition, there are places where street maps are not commonly used and instead landmarks are used allowing the intuitive navigation by recognizable and memoizable views along the route. The introduction of buildings as landmarks together with corresponding spoken instructions is a step towards a more natural navigation. The integration of GPS and GIS provides this capability. The main problem lies in identifying suitable landmarks and evaluating their usefulness for navigation instructions. Existing databases can help to tackle this problem and be an integrated part of most navigation applications. For example, Brondeel et al. [2] used GPS, GIS, and accelerometer data to collect data of trips and proposed a prediction model for transportation modes with high correction rate. ResZexue et al. [3] developed a logistics distribution manager (LDM) software and a smart machine (SM) system. It is based on fusing GPS, GIS, Big Data, Internet+, and other technologies to effectively apply its attributes and benefits for achieving a robust information management system for the logistics industry. The resulting logistics facility has shorter distribution time, improved operational competitiveness, optimized the workflow of the logistics distribution efficiency, and saved cost. These examples demonstrate the level of improvements we can expect by integrating GPS and GIS as well as the IoT, mobile phones, and other current technologies.

3. GPS and mobile phone integration

GPS positions provided via phone are generated using multiple different methods, resulting in highly variable performance. Performance depends on the smartphone attributes, the cell network, availability of GPS satellites, and line of sight to these satellites. The time from turning on the smartphone to getting GPS coordinates is relatively long. To accelerate it, a variety of techniques got used. Some phones have incomplete GPS hardware, requiring a cell network to function. The quality of the GPS antenna determines the duration until the device will get a lock. For example, the S3 Mini device has relatively good GPS hardware, including GLONASS and A-GPS support.

4. Urban vehicles navigation

Urban canyons, sky blockage, and multipath errors affect the quality and accuracy of GNSS/GPS. Public transportation in modern cities may have hundreds of routes and thousands of bus stops, exchange points, and busses. These two factors make urban bus systems hard to follow and complex to navigate. Mobile applications provide passengers with transport planning tools and find the optimal route, next bus number, arrival time, and ride duration. More advanced applications provide also micro-navigation-based decisions, such as current position and bus number, the number of stops left till arrival, and exchange to a better route. Micro-navigation decisions are highly contextual and depend not just on time and location but also on the user's current transport mode, waiting for a bus or riding on a bus. Emerging technology is where accuracy and robustness are critical requirements for safe guidance and stable control. GNSS accuracy can be significantly improved using several techniques such as differential GNSS (DGNSS), augmented GNSS, and precise positioning services (PPS). These techniques add complexity and additional cost. Multi-constellation GNSS also enhances the accuracy by increasing the number of visible satellites. In dense urban areas where high buildings are common, the geometry of visible satellites often results into high uncertainty in the vehicle's GNSS position estimate resulting in performance in dense urban areas still being challenging.

4.1 Bus navigation using embedded Wi-Fi and a smartphone application

Urban Bus Navigator (UBN) is a system infrastructure that connects passengers' mobile smartphones with Wi-Fi-enabled busses, gaining real-time information about the journey and transport situations of passengers [4]. A key feature of UBN is a semantic bus ride detection that identifies the concrete bus and route the passenger is riding on, providing continuous, just-in-time dynamic rerouting and end-to-end guidance for bus passengers. Technical tests indicate the feasibility of semantic bus ride detection, while user tests revealed recommendations for effective user support with micro-navigation. The system elements include semantic bus ride detection using a Wi-Fi-based recognition system and a dynamic trip tracking. The semantic bus ride detection combined with the phone's GPS is used to monitor the passenger's trip progress. Deviations are immediately recognized and trigger replanning the trip, resulting a new set of navigation instructions for the passenger. The architecture is composed of Wi-Fi for proximity detection of busses by the passenger's mobile phone, a smartphone application for trip planning using macro-navigation, a context-aware trip hints using micro-navigation, context sensing, bus ride recognition, and trip tracking.

4.2 GNSS/IMU sensor fusion scheme

This urban navigation is based on detecting and mitigating GNSS errors caused by condensed high buildings interfering signals going through [5]. It is using a map-aided adaptive fusion scheme. The method estimates the current active map segment using dead-reckoning and robust map-matching algorithms modeling the vehicle state history, road geometry, and map topology in a hidden Markov model

(HMM). The Viterbi algorithm decodes the HMM model and selects the most likely map segment. The projection of vehicle states onto the map segment is used as a supplementary position update to the integration filter. The solution framework has been developed and tested on a land-based vehicular platform. The results show a reliably mitigate biased GNSS position and accurate map segment selection in complex intersections, forks, and joins. In contrast to common existing adaptive Kalman filter methods, this solution does not depend on redundant pseudo-ranges and residuals, which makes it suitable for use with arbitrary noise characteristics and varied integration schemes.

4.3 Navigation based on compass-based navigation control law

Urban environments offer a challenging scenario for autonomous driving [6]. The proposed solution allows autonomously navigate urban roadways with minimum a priori map or GPS. Localization is achieved by Kalman filter extended with odometry, compass, and sparse landmark measurement updates. Navigation is accomplished by a compass-based navigation control law. Experiments validate simulated results and demonstrate that, for given conditions, an expected range can be found for a given success rate.

The architecture contains steering and speed controllers, an object tracker, a path generator, a pose estimator, and a navigation algorithm using sensors allowing real-time control. High-level localization is provided by the pose estimator, which utilizes only odometry measurements, compass measurements, and sparse map-based measurements. The sparse map-based measurements generated from computer vision methods compare raw camera images to landmark images contained within a sparse map. The roadway scene includes lane line markings, road signs, traffic lights, and other sensor measurements. The scene information and the inertial pose estimate are fed into a navigation algorithm to determine the best route required to reach the target. This navigation scheme is provided by a compass-based navigation control law.

5. Space navigation systems

Common navigation technologies assume navigation on a surface with two-dimension (2D), flat land area. Navigation in three-dimension (3D) is much more complicated requiring at least new technologies to complement the existing 2D navigation technologies.

5.1 Autonomous navigation of micro aerial vehicles

In this section we present a low-computational method for state estimator enabling autonomous flight of micro aerial vehicles [7]. All the estimation and control tasks are solved on board and in real time on a simple computational unit. The state estimator fuses observations from an inertial measurement unit, an optical flow smart camera, and a time-of-flight range sensor. The smart camera provides optical flow measurements and odometry estimation, avoiding the need for image processing, usable during flight times of several minutes. A nonlinear controller operating in the special Euclidean group SE(3) can drive, based on the estimated vehicle's state, a quadrotor platform in 3D space guaranteeing the asymptotic stability of 3D position and heading. The approach is validated through simulations and experimental result.

5.2 Vision-based navigation for micro helicopters

Weiss [8] developed a vision-based navigation system for micro helicopters operating in large and unknown environments. It is based on vision-based methods and a sensor fusion approach for state estimation and sensor self-calibration of sensors and with their different availability during flight. This is enabled by an onboard camera, real-time motion sensor, and vision algorithms. It renders the camera and an onboard multi-sensor fusion framework capable to estimate at the same time the vehicle's pose and the inter-sensor calibration for continuous operation. It runs at linear time to the number of key frames captured in a previously visited area. To maintain constant computational complexity, improve performance, and increase scalability and reliability, the computationally expensive vision part is replaced by the final calculated camera pose.

5.3 Space navigation using formation flying tiny satellites

Traditional space positioning and navigation are based on large satellites flying in a semi-fixed orbit and so are costly and less flexible [9]. Recent developments of low-mass, low-power navigation sensors and the popularity of smaller satellites, a new approach of having many tiny spacecrafts flying in clusters under controlled configurations utilizing its cumulative power to perform necessary assignments. To keep stable but changeable configurations, positioning, attitude, and intersatellite navigation are used. For the determination of relative position and attitude between the formation flying satellites, Carrier-phase differential GPS (CDGPS) is used, where range coefficients, GPS differential corrections, and other data are exchanged among spacecrafts, enhancing the precision of the ranging and navigation functions. The CDGPS communicates the NAVSTAR GPS constellation to provide precise measures of the relative attitude, the positions between vehicles, and attitude in the formation.

6. Pedestrian navigation systems

Pedestrian navigation services enable people to retrieve precise instructions to reach a specific location. As the spatial behavior of people on foot differs in many ways from the driver's performance, common concepts for car navigation services are not suitable for pedestrian navigation. Cars use paved roads with clear borderlines and road signs, and so keeping the car on track is its main role, neglecting obstacles and hazards, unless it is integrated with a social network. However, pedestrians, unlike like cars, may not follow the defined road. This makes personal navigation more complicated and forces us adding special features required for safe navigation. Pedestrian navigation requires very accurate, high-resolution, and real-time response [10]. Solely GPS does not support last moment route changes, such as road detours, significant obstacles, and safety requirements. However, integrating the IoT and GPS via an application generates a solution providing accurate and safe navigation. To enable it, a two-stage personal navigation system is used. In the first stage, the trail is photographed by a navigated drown, and the resulting video is saved in a cloud database. In the second stage, a mobile application is loaded to the pedestrian's mobile phone. Once the pedestrian is about to walk, it activates the mobile application which synchronizes itself with the cloud navigation database, and then instructions from the mobile phone guide the pedestrian along the trail-walk. A more advanced system contains the two stages within the mobile application. The mobile video camera is activated and captures the trail images in front of the pedestrian, processes it, and guides the pedestrian accordingly. In case of an upcoming obstacle, the application proposes the safest and most effective detour and guides the pedestrian accordingly.

Personal navigation systems are very accurate and safe, operate indoor and outdoor, and are available as long as the mobile phone is connected, and its internal storage is big enough. It provides spatial information for climbing, wandering, or tramping users. It is used for locating casualties, as well as for self-orientation of rescue teams in areas with low visibility. In military and security operations, localization and information technologies are used by soldiers to self-locate, collect, and collate. A similar implementation with the same functionality is a walking stick with embedded micro devices and software as described above and a wearable Bluetooth headset with an embedded camera in front of it.

6.1 Landmark-based pedestrian navigation systems

Navigation in cities is commonly done by the target address: zip-code, street, and house number. However, in cases where people do not use street and house number as an address but rather use landmarks to identify the route to the target as well as the target location [11], by combining CIS and GPS, the desired landmarks coordinates are loaded to the cloud database, and the corresponding navigation application is modified to identify the landmarks on ground.

A landmark-based navigation system is composed of a video camera to obtain and analyze pedestrian paths, selected reliable landmarks along the main routes, a routing table containing all relevant origins and destinations within the site, positions of view and orientations to assert maximal coverage of interesting spots, thousands of partial routes for the entire recording period, and the detected stops over a whole day for different definitions of a stop. Based on the defined sections and the landmarks and decision points, a routing table is created to define navigational instructions from each origin in the station to each possible destination. Table columns correspond to the original landmarks and the decision points; rows correspond to destination landmarks. The identified landmarks and the defined route instructions are used to develop an audio guiding system using speech recognition and text-to-speech software. The audio guiding system employs verbalisms that are as distinct and clearly recognizable as the visual landmarks and that the users can intuitively combine the description with what they see.

6.2 Shoe navigation based on micro electrical mechanical system

A micro electrical mechanical system (MEMS) [12] is a family of thumbnail technologies enabling a wide variety of advanced and innovative applications. When such device is mounted on a shoe, it collects the number of steps, average step width, and walking directions. This data is constantly collected and processed, and via signals it guides the person wearing the shoe. Due to the magnetic field, some navigation errors may occur; a special filter offsets it by using a special filter. Experiments show that this approach is applicable and efficient.

7. Indoor navigation technologies

Indoor navigation systems became popular due to the lack of GPS signals indoors and the increase in navigation needs especially in small areas, such as parking garages and huge complex of buildings. Several indoors navigation systems have already been implemented. Each of them is based on a different technology that complies with the specific requirements and constraints of the location it is expected to navigate in. We assume that each solution has technical and usability limitations. It helps tracking objects by using wireless concepts, optical tracking, ultrasound techniques, sensors, infrared (IR), ultra-wide band (UWB), Wireless Local Area Networks (WLANs), Wi-Fi, Bluetooth, radio frequency identification (RFID), assisted GPS (A-GPS), and more. Most solutions have limited capabilities, accuracy, unreliability, design complexity, low security, and high configuration costs.

7.1 NFC-based indoor navigation system

NFC technology allows communication over short-range, mobile, and wireless conditions. NFC communication happens when two NFC-capable devices are close to each other. Users use their NFC mobiles to interact with an NFC tag or another NFC mobile. NFC-based indoor navigation system enables users to navigate through a complex of buildings by touching NFC tags spread around and orienting users to the destination [13]. NFC internal has considerable advantages to indoor navigation systems in terms of security, privacy, cost, performance, robustness, complexity, and commercial availability. The application orients the user by receiving the destination name and touching the mobile device to the NFC tags and so navigates to the desired destination.

7.2 Indoor garage navigation based on car-to-infrastructure communication

Indoor micro-navigation systems for enclosed parking garages [14] are based on car-to-infrastructure communication providing layout information of the car park and the coordinates of the destination parking lot. It uses unique signal rates. In case a car is detected, the system calculates its position and transmits data to a vehicle to substitute the internal positioning system. With this information the vehicle is guided. Integration to the outdoor navigation system is available to allow smooth transition from/to outdoor/indoor.

7.3 Autonomous vision-based micro air vehicle (MAV) for indoor and outdoor navigation

In this section we introduce a quadrotor that performs autonomous navigation in complex indoor and outdoor environments [15]. An operator selects target positions in the onboard map, and the system autonomously plans flights to these locations. An onboard stereo camera and an inertial measurement unit (IMU) are the only sensors. The system is independent of external navigation aids like GPS. All navigation tasks are implemented onboard the system. The system is based on FPGA-dense stereo matching images using semi-global matching, locally drift-free visual odometry with key frames and sensor data fusion. It utilizes the available depth images from stereo matching. To save processing time and make large movements or rather low frame rates possible, the system works only on features. A wireless connection is used for sending images and a 3D map to the operator and to receive target locations. The results of a complex, autonomous indoor/outdoor flight support this approach. The position is controlled by the estimated motion of the sensor. To enable it, a state machine controller, a tracking position system, and a reference generator are implemented. The reference generator is used to create smooth position, velocity, acceleration, and a tracking controller based on a list of waypoints. The flown path is composed of straight line segments between any two waypoints.

8. Obstacle avoidance navigation systems

A comprehensive automated navigation system must incorporate effective tools for detecting road obstacles and instantly propose the optimal alternate route bypassing the detected obstacle. It combines optimal route finding, real-time route inspection, and route adjustments to ensure safe navigation. The following are three examples utilizing advanced technologies such as computer vision, fuzzy logic, and context-aware. More examples can be found in [16].

8.1 Image processing obstacle avoidance navigation

Unmanned aerial vehicles (UAVs) use vision as the principal [17] source of information through the monocular onboard camera. The system compares the obtained image to the obstacles to be avoided. Micro aerial vehicle (MAV), to detect and avoid obstacles in an unknown controlled environment. Only the feature points are compared with the same type of contrast, achieving a lower computational cost without reducing the descriptor performance. After detecting the obstacle, the vehicle should recover the path. The algorithm starts when the vehicle is closer to the obstacle than the distance allowed. The limit area value is experimentally obtained defining the dimensions of obstacles in pixels at a specific distance. The output of the control law moves the vehicle away from the center of the obstacle avoiding it. If the error is less than zero, the vehicle moves to the right side. Detouring of permanent obstacles, a preliminary process is applied to scan the route and correct it such that the corrected route already considers all known obstacles and skips them.

8.2 Fuzzy logic technique for mobile robot obstacle avoidance navigation

Mobile robots perform tasks such as rescue and patrolling. It can navigate intelligently by using sensor control techniques [18]. Several techniques have been applied for robot navigation and obstacle avoidance. Fuzzy logic technique is inspired by human perception-based reasoning. It has been applied to behavior-based robot navigation and obstacle avoidance in unknown environments. It trains the robot to navigate by receiving the obstacle distance from a group of sensors. A reinforcement learning method and a genetic algorithm optimize the fuzzy controller for improving its performance while the robot moves. Comparing the performance of different functions such as triangular, trapezoidal, and Gaussian for mobile robot navigation shows that the Gaussian membership function is more efficient for navigation.

A similar concept is using neural network learning method to construct a path planning and collision-free for robots. Real-time collision-free path planning is more difficult when the robot is moving in a dynamic and unstructured environment.

8.3 Context-aware mobile wearable system with obstacle avoidance

The system is composed of three embedded components; a map manager, a motion tracker, and a hindrance dodging [19]. The map manager generates semantic maps from a given building model. The hindrance dodging detects visible objects lying on the road and suggests a safe bypass route to the target location. A developed prototype performed very well proving that this navigation system is effective and efficient.

9. Conclusions

This chapter introduces various complementing navigation concepts and implementations, integrating advanced technologies and improving and expanding existing traditional navigation solutions. The outcome is a wide variety of solutions for cases where standard navigation technologies such as GPS are less effective or not applicable. We presented several areas where various technologies have been tailored to specific problems. For each problem we described different cases with unique technologies and implementations.

Author details

Menachem Domb
Ashkelon Academic College (AAC), Israel

*Address all correspondence to: dombmnc@edu.aac.ac.il

References

[1] Taczanowska K, Muhar A, Brandenburg C. Potential and limitations of GPS tracking for monitoring spatial and temporal aspects of visitor behavior in recreational areas. In: Raschi A, Trampetti S, editors. Management for Protection and Sustainable Development – Proceedings of the Fourth International Conference on Monitoring and Management of Visitor Flows in Recreational and Protected Areas, Montecatini Terme, 14-19 October, 2008; 2008. pp. 451-455

[2] Ruben B, Pannier B, Chaix B. Using GPS, GIS, and accelerometer data to predict transportation modes. Medicine and Science in Sports and Exercise. 2015

[3] Li R, Ding Z, Qin Q, Huang H, Guo Y. Research on logistics route optimization based on GPS and GIS technology. In: 3rd International Conference on Electronic Information Technology and Intellectualization, ICEITI 2017; 2017

[4] Foell S, Kortuem G, Rawassizadeh R, Handte M, Iqbal U, Marrón P. Micro-navigation for urban bus passengers: using the Internet of things to improve the public transport experience. In: Proceedings of the First International Conference on IoT in Urban Space, ACM 2014; 2014. pp. 01-6

[5] Mohamed M. Atia a,⇑ , Steven L. Waslander, map-aided adaptive GNSS/IMU sensor fusion scheme for robust urban navigation. Measurement. 2019;**131**:615-627

[6] Chipka JB , Campbell M. Autonomous urban localization and navigation with limited information. In: 2018 IEEE Intelligent Vehicles Symposium (IV); 2018. pp. 7-12

[7] Santamaria-Navarro A, Loianno G, Sol'a J, Vijay Kumar J, Andrade-Cetto. Autonomous navigation of micro aerial vehicles using high-rate and low-cost sensors. Autonomous Robots. August 2018;**42**(6):1263-1280

[8] Weiss SM. Vision Based Navigation for Micro Helicopters. Available from: https://doi.org/10.3929/ethz-a-007344020, ETH 2012

[9] Zencik RG, Kohlhepp K. GPS micro navigation and communication system for clusters of micro and nanosatellites. In: 2001 IEEE Aerospace Conference Proceedings (Cat. No.01TH8542); 2001

[10] Millonig A, Schechtner K. Developing landmark-based pedestrian navigation systems. IEEE Transactions on Intelligent Transportation Systems. 2007;**8**(1):43-49. DOI: 10.1109/TITS.2006.889439

[11] Millonig A, Schechtner K. Developing landmark-based pedestrian navigation systems. IEEE Transactions on Intelligent Transportation Systems. 2007;**8**(1). DOI: 10.1109/TITS.2006.889439

[12] Cho SY. MEMS based pedestrian navigation system. Journal of Navigation. 2006:135-153. DOI: 10.1017/S0373463305003486

[13] Ozdenizci B, Coskun V, Ok K. NFC internal: An indoor navigation system. Sensors. 2015;**15**:7571-7595. DOI: 10.3390/s150407571

[14] Einsiedler J, Sawade O, Schäufele B, Witzke M, Radusch I. Indoor micro navigation utilizing local infrastructure-based positioning. In: July 2012 IEEE Intelligent Vehicles Symposium; 2012

[15] Schmid K, Lutz P, Tomic T, Mair E, Hirschmüller H. Autonomous vision-based micro air vehicle for indoor and outdoor navigation. Journal of Field Robotics. 2014;**31**(4, Special Issue on

Low Altitude Flight of UAVs):537-570. DOI: 10.1002/rob.21506

[16] Polvara R, Sharma S, Sutton R, Wan J, Andrew. Manning obstacle avoidance approaches for autonomous navigation of unmanned surface vehicles. Journal of Navigation. 1-16. DOI: 10.1017/S0373463317000753

[17] Aguilar WG, Casaliglla VP, Pólit JL. Obstacle avoidance based-visual navigation for micro aerial vehicles. Electronics. 2017;**6**(1):10. DOI: 10.3390/electronics6010010

[18] Pandey A, Pandey S, Parhi DR. Robot navigation and obstacle avoidance techniques: A review. International Robotics and Automation Journal. 2017;**2**(3). DOI: 10.15406/iratj.2017.02.00023, 2017

[19] Li B, Munoz JP, Rong X, Xiao J, Tian Y, Arditi A. ISANA: Wearable Context-Aware Indoor Assistive Navigation with Obstacle Avoidance for the Blind, European Conference on Computer Vision, ECCV 2016: Computer Vision – ECCV 2016 Workshops; 2016. pp. 448-462

Chapter 3

A Review of the Machine Learning in GIS for Megacities Application

Nasim Tohidi and Rustam B. Rustamov

Abstract

Machine learning (ML) is very useful for analyzing data in many domains, including the satellite images processing. In the remote sensing data processing, ML tools are mainly founded out a place for filtering, interpretation and prediction information. Filtering aims at removing noise and performing transformations, which is vital segment of data processing as useful performance of data validation. An interpretation is significant part of it as the stage of objects classification depends of existing task for solution. Prediction is performed to estimate precise values of underlying parameters or future events in the data. It can be used successfully above achievements in a variety of areas. An urbanization is one of the spheres of advance technology application where highly need to collect appropriate data for understanding of challenges facing society. The process of urbanization becomes very important problem, thanks to city expansion. Each city is a complicated system. It consists of various interactive sub-systems and is affected by multiple factors, including population growth, transportation and management policies. To understand the driving forces of the urban structure change, the satellite-based estimates are considered to monitor these changes, in long term. GIS (geographic information system) is equivalent to methods related to the use of geospatial information. Besides, the increasing application of ML techniques in various fields, including GIS, is undeniable. Thus, the chapter attempts to review the application of ML techniques in GIS with a focus on megacities and theirs features fixing/identification and solution.

Keywords: geographic information system, machine learning, urbanization, data processing, modeling

1. Introduction

Today there is a growing need for the collection, processing, management and efficiently use of reliable spatial information. Therefore, it is very significant to be aware of relevant approaches and to share experiences and develop best practices. This growing demand is due to the most important developments in society, which in turn are magnified by rapid urbanization and the conditions of the megacities.

Location, in the form of spatial data, is a key point for visualizing the current location, predicting events and enhancing service delivery. Information about location can integrate and strengthen the complex analysis of the distribution of locations, events, and services. This provides many opportunities for improving government services in terms of best governmental segments, interacting with customers and optimizing processes. As cities get larger, spatial information becomes like a key tool in efficient urban service delivery, public safety, and overall resource management.

On the other hand, today, artificial intelligence methods, especially ML techniques, have come to the attention of scientists and officials in various fields, to analyze and manage the enormous data that is produced at any given moment, and one of the most exciting tools that have entered the material science toolbox in recent years is ML. Undoubtedly, one of these fields is GIS.

In practice, a GIS allows users to understand the spatial dimensions of their work and relate it to information such as population information as well. The data collected and stored by the GIS can be used for different purposes ranging from transport, draught analysis, agriculture, disease-outbreak analysis, land occupancy, etc. At the same time GIS makes possible to storage a big volume of data in safely stage and access to them at any needed time and rapid base. So, the goal of this chapter is to review past works and research in this area, because it can be supposed that can help greatly in understanding the current situation and capabilities; besides, it will be attempt of step in planning for future developments in the field of GIS.

The remainder of this chapter is organized and structures as follows. In Section 2, main definitions are mentioned. Section 3 presents an overview of ML application in GISs and related works in this area. In Section 4, it has been introduced the evaluation metrics and datasets. The last Section 5 provides conclusions.

2. Fundamental principles

In order to review the ML application in GIS, the first is needed to familiarize with the basic concepts in this regard. The followings are some fundamental principles and definitions.

2.1 Machine learning

ML is an application of artificial intelligence that provides systems the ability to automatically learn and improve their performance from experience without being explicitly programmed. ML focuses on the development of computer programs that can access data and use it in the process of learning [1].

The process of learning begins with observations or data, such as examples, direct experience, or instruction. The data will be used in order to look for patterns in it and make better decisions based on the provided examples. The primary goal is to allow the machines learn without human intervention or assistance and adjust actions accordingly.

ML algorithms are often categorized as supervised or unsupervised, however this categorization is very general and it cannot cover all of the available methods [2]:

- Supervised ML algorithms can use what has been learned in the past by using labeled examples to predict future events from unseen data. Starting from the analysis of a training dataset (labeled examples), the learning algorithm predicts the output values. The system is capable of providing targets for each new input after sufficient training. Besides, the algorithm can compare its output with the correct output and find errors to modify the model accordingly. Examples of these algorithms: Support Vector Machine (SVM), Decision Tree, Random Forest, KNN, Regression, etc.

- Unsupervised ML algorithms are used when the training data is not labeled or classified. The purpose of these algorithms is to examine how systems can derive a function to describe the hidden pattern of unlabeled data. They may not specify the appropriate output, but it explores data and can infer to describe hidden structures from unlabeled data. Examples of these Learning: Apriori algorithm, K-means, EM.

- Semi-supervised ML algorithms fall in between the two types of previously mentioned algorithms, because they use both labeled and unlabeled data for training. Usually, a small portion of data is labeled and a large amount of it, is unlabeled. The systems that use these algorithms can achieve high level of accuracy. Typically, semi-supervised learning is selected when the acquired labeled data requires skilled and relevant resources in order to learn from it (producing labeled data costs money and takes time.). Otherwise, accessing to unlabeled data generally does not require additional resources.

- Reinforcement learning algorithms are learning methods that interacts with its environment by generating actions and receiving punishments or rewards. Trial and error search and delayed reward are the most important features of these algorithms. They allow systems agents to automatically determine the ideal behavior in a particular context in order to maximize its performance quality. Simple reward feedback is known as the reinforcement signal. Examples of these Learning: Q-learning, Markov Decision Process.

ML enables analysis of massive amount of data. Besides, it generally provides faster, more accurate results in order to identify profitable opportunities or dangerous risks, it may also require additional time and resources to train it properly. ML requires formatted data that is analyzed to build a ML model.
In other words, it requires an appropriate set of data that can be applied to a learning process.

ML can be used in cases where using human resources is not time/cost effective or when many variables are being considered simultaneously. ML uses the prepared data to train a ML algorithm. An algorithm is a computerized procedure or recipe. When the algorithm is trained on the data, a ML model will be generated. Once the data is prepared and the algorithm trained, the ML model can make predictions about the unseen data, on its own.

Selecting the right algorithm for the issue is necessary for applying ML successfully. Selection is largely influenced by the application and the data available.

2.1.1 Choosing the most appropriate ML algorithm

There are a large number of ML algorithms available. Choosing the optimal algorithm for a specific problem is dependent on its features such as speed, accuracy, training and predicting time, amount of data required to train, data type, how easy is it to implement, etc. Most of the time, for GIS applications, time is very important.

To avoid dependence on the specific conditions, it is common to analyze the runtime of algorithms in an asymptotic sense. So, considering n the number of training sample, p the number of features, n_t the number of trees, n_{sv}, the number of support vectors and k, the number of clusters, following are time complexity factors of some ML algorithms, which help to choose the correct algorithm for the issue (**Table 1**):

Algorithm	Learning	Predicting
Regression	$O(p^2n + p^3)$	$O(p)$
Decision Tree	$O(n^2p)$	$O(p)$
Random Forest	$O(n^2pn_t)$	$O(pn_t)$
Naïve Bayes	$O(np)$	$O(p)$
SVM	$O(n^2p + n^3)$	$O(pn_{sv})$
KNN	—	$O(np)$
K-means	$O(n^{pk+1})$	$O(k)$

Table 1.
Time complexity of some ML algorithms.

Where:

- Time for Learning is time associate with training of dataset. It varies with size of data and algorithm we are using in that.
- Time for Predicting is time associate with testing of dataset or predicting unseen data. It varies with size of data and algorithm we are using in that.

Most of the time, about 80 percent of the dataset will be used for training and the remaining part will be used for tuning and testing. In addition, it should be noted that, as the training phase most of the time can be performed offline, the predicting time is more important for developers.

Generally, it can be used the points above to shortlist a few algorithms, but it is hard to know right at the start which algorithm will work in the best way. It is usually desirable to work iteratively. Among the ML algorithms can be identified as potential good approaches, throw the data into them, run them all in either parallel or serial, and at the end evaluate the performance of the algorithms to select the best one(s).

2.2 Megacity

A megacity is defined by the United Nations (UN) as a city which has a population of 10 million or more people. Currently, there are 38 megacities in the World (**Figure 1**). The UN statistics indicate that the city with the largest populations worldwide is Tokyo with 38.8 million people. Recently, the UN has predicted that the number of megacities will rise to 41 by the year 2030.

The urbanization process poses enormous challenges for governments, social and environmental planners, engineers, architects and the residents of the megacities. No wonder, the growing population of cities creates demand, in areas such as housing and services. The environmental destruction and poverty are two other concerns, which city administrations have to take care of, as especially poor people do not have the necessary financial resources to tackle these problems.

Megacities affect a variety of living conditions for citizens. Although stress level, traffic jams, poor air quality and increasing health risks, make life more difficult in megacities, most people still choose to live there. Therefore, more accurate governmental programs are needed to help improve living conditions for the metropolitan inhabitants.

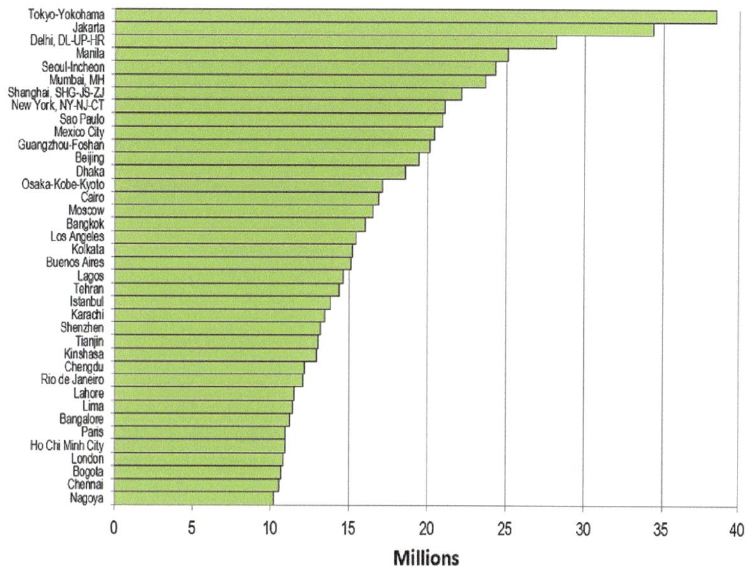

Figure 1.
The 38 megacities in 2019 [3].

As more cities are becoming megacities and existing megacities are growing, policymakers and urban planners are grappling with the questions of how to make growth at this scale sustainable, and how to tackle the escalating social, economic and environmental problems evident in the world's megacities. One of the most popular solutions is ML.

3. Application of machine learning in GIS

Urban dispersal and expansion has become an important issue for municipalities, environmental scientists and urban planners. Especially, in megacities, this issue becomes more vital. Currently, more than %50 of the world's population lives in urban areas and then it is predicted to grow over the %65 by 2050, according to the United Nation report. For example, all population in the 500,000+ urban areas of Australia and New Zealand combines to equal that of Moscow or Bangkok, and only slightly larger than Los Angeles (16.4 million). It is known that developing countries have already begun a rapid urbanization [4]. The fact that the global population has increased rapidly since the industrial revolution of the 18th century, highlights the problems of urban planning and urbanization, because of the population gathering in certain centers [5]. This unnatural pace of urbanization has created significant social and environmental challenges for decision-makers [6]. In addition, modeling and simulation are effective tools for discovering the urban development mechanisms and for providing planning in growth management. Therefore, monitoring and modeling the urban sprawl of cities is a necessary key parameter to prevent precautions [7, 8].

As it has been illustrated in **Figure 2**, Asia remains the dominance in terms of megacities, with nearly 58 percent of the population in larger metropolitan areas. This is approximately five times as many greater urban area residents as in North America or Africa. Besides, Asia has more than five times as many larger urban area residents as Europe and eight times that of South America [3].

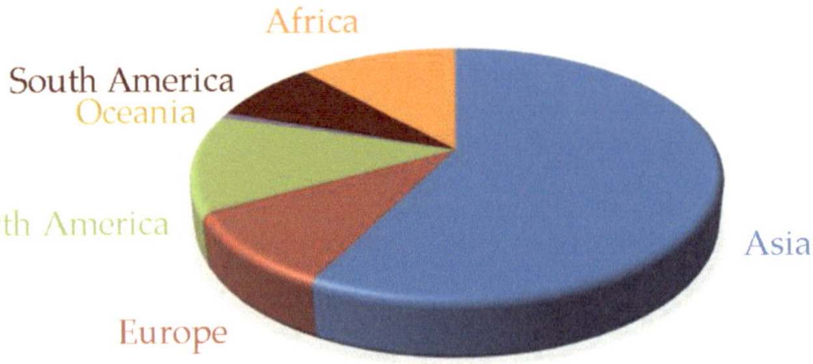

Figure 2.
Built-up urban area population in 2019.

Urban expansion modeling became widespread in the 1960s [9] and have accelerated with the developments of the technologies like Remote Sensing (RS) and GIS. Today, RS has been widely recognized as an essential tool for urban planning, management and design due to allows to get spatiotemporal data that are necessary for modeling environmental impacts, urban expansion and population growth. Particularly, the benefits of satellite-based image data have attracted attention of the scientific studies on urban expansion and environment [10, 11]. RS enables the collection of spatial details data for large areas at different time intervals; therefore, it provides a unique perspective on revealing the spatial and temporal dynamics of the change process in the land use and urban expansion [12]. GIS technology is described as an effective tool for identifying and monitoring the land cover change at different scales [13–16].

The dynamic modeling via GIS as a tool for urban simulation has rapidly gained popularity [7, 17]. The application of ML models has increased noticeably in RS filed due to the increased availability of powerful and flexible ML software and improvements in computing performance [18, 19].

The useful application of artificial neural networks (ANNs) in interpreting spatial resource information have been proven i.e. one of the most common are back-propagation neural networks, which are widely used by spatial planners. However, to improve the usability of ANNs for map-based applications, a more efficient method for communicating between the GIS and a trained ANN, is critical. When ANNs and GIS are used together for many applications to improve decision-making quality. The ANN design will consist of numerous layers, all of which can have different weights. The training process of an ANN involves changing the weightings over time until as it is desired the network reaches the static or optimum firing state. After training, an ANN can be used for applications effectively and consistently. Through the application of an ANN, GIS professionals can add another dimension to their spatial capabilities. In some research, a combination of neural networks with remote sensing image data for mapping the urbanization dynamics, has been proposed [20–22].

As another helpful technic, recently evolutionary algorithms have been used to tackle a variety of complex computation and optimization problems, such as natural language processing [23], route finding [24] and image processing [25]. One of the most important applications of evolutionary algorithms is in the field of GIS [26–28]. It should be noted that generally these algorithms are not fast in comparison with other ML algorithms and their main usage is for optimization.

All in all, there are two main ways for satellite-image processing that each of them has its own advantages and disadvantages. Sometimes there is high-resolution data (satellite images) so processing this amount of data would take a lot of time, with high accuracy. Some other cases, depending on the issue the resolution is not good, but the measurement methods are highly qualified.

3.1 Process description: impact and influence determination

Over the years, the fast development of map services [29] and volunteered geographic information (VGI) [30] has provided a huge number of geo-tagged images. This data source has given information on every corner of a city and has been enabling broader and more in-depth quantitative research in related fields. These data improve the understanding of the dynamic and physical features of the city by identifying landmark [31], detecting urban identities [32, 33], assessing the living environment inequality [34], and modeling human activities [35] and popular places [36]. Also, they provide information on the social and physical structures of dynamic metropolitan environments [36, 37].

The MIT Media Lab launched the "Place Pulse" program in 2013, which is a data collection platform that enables volunteers to take part in the urban perception rating experiment. By the end of 2016, the MIT Place Pulse dataset had collected 1,170,000 pairwise comparisons from 81,630 online participants for 110,988 cityscape images. Given this dataset and advances in ML techniques, many studies have tried to analyze human perceptions of urban appearance [34, 38–41].

Since previous approaches use low or mid-level image attributes, they have problem in extracting high-level information about the natural image. Some examples of these attributes are: Gist, SIFT- Fisher Vectors, DeCAF features [41], geometric classification map, color Histograms, HOG2x2, and Dense SIFT [40].

For instance, according to building models, SVM and Linear Regression were used in [41] to predict image labels. Support Vector Regression was used in [40], Ranking SVM it has been used in [42], and several convolutional neural network (CNN) based approaches were used in [39, 41, 42].

Among the various image representations and models, approaches that uses deep convolutional neural network (DCNN), have outperformed conventional methods to a large extent. In [39], authors introduce a DCNN model that is based on the Deep Residual Network (ResNet) [43], which won the first place in the ImageNet Large Scale Visual Recognition Competition [44].

Recently, a shift-invariant and hierarchical model has emerged in the form of DCNN, because of the availability of large-scale annotated datasets and the rapid development of high-performance computing systems. In some research, DCNN was employed to conduct human perception modeling and prediction. Due to its powerful capability to learn and represent automatic image feature, this model has attracted a lot of attention and achieved great success in multiple fields, including speech recognition [45], natural language processing [46], and visual object detection [43, 47].

A very deep convolutional neural network is hard to train and optimize because of disappearance gradients and the curse of dimensionality [43, 48]. ResNet is known as an acceptable attempt to address this problem. It was designed to learn the residual functions with regard to the layer inputs rather than learning the unreferenced functions [43].

In [49], a data-driven ML approach that measures how people perceive a place in a large-scale urban region, was proposed. In particular, a deep learning model, that had been trained with millions of human ratings of street-level imagery, was used to predict human perceptions of a street view image. The model achieved a high accuracy rate in predicting six human perceptual indicators. These indicators are: beautiful, boring, depressing, lively, safe and wealthy. It could help to map the distribution of the city-wide human perception for a new urban area. Besides, to determine the visual points that may cause a place to be perceived as various perceptions, a series of statistical analyses was performed. From the 150 object categories that had been segmented from the images of the street view, many objects were identified as being positively or negatively correlated with each of the indicators. The mentioned results helped urban planners and researchers to take a step toward getting the interactions of the place sentiments and semantics.

Big data are voluminous and complex data of different qualities, that have the potential to generate new hypotheses and new methods for understanding interactions between social, biophysical and infrastructure domains of complex urban systems that face the challenges of climate change [50, 51].

The movement of people tracked through cell phones is an example of crowd-sourced and big data, which offers manifold new possibilities for assessing the city's inner workings, and the availability, quality and quantity of data is evolving, rapidly (**Figure 3**). Crowd-sourced information can be used as a reliable proxy, with much better resolution and replication, for more traditional methods of empirical social survey [53].

Similar analyses of social media provide the opportunity to complement the existing traditional ways of collecting information about human behavior in cities, which can be brought together with other sources of biophysical and infrastructural data, especially in spatial formats, collected through GIS [54, 55].

Big data can also come from urban hotlines, city planning offices, tax assessor databases, records about utility use and repair, and the rapid emergence of sensors and instrumented buildings, ecological spaces and even roads [56].

In **Figure 3**, The direction of change is shown by color, where I equals warmer and wetter; II colder and wetter; III colder and drier; and IV warmer and drier conditions. The direction of change is measured with the Euclidean distance in the 2D space including temperature and precipitation change. The classification of the magnitude of change corresponds to quartiles.

The usefulness of big data for understanding urban systems such as efficacy of climate solutions and climate impacts will only increase with time [56].

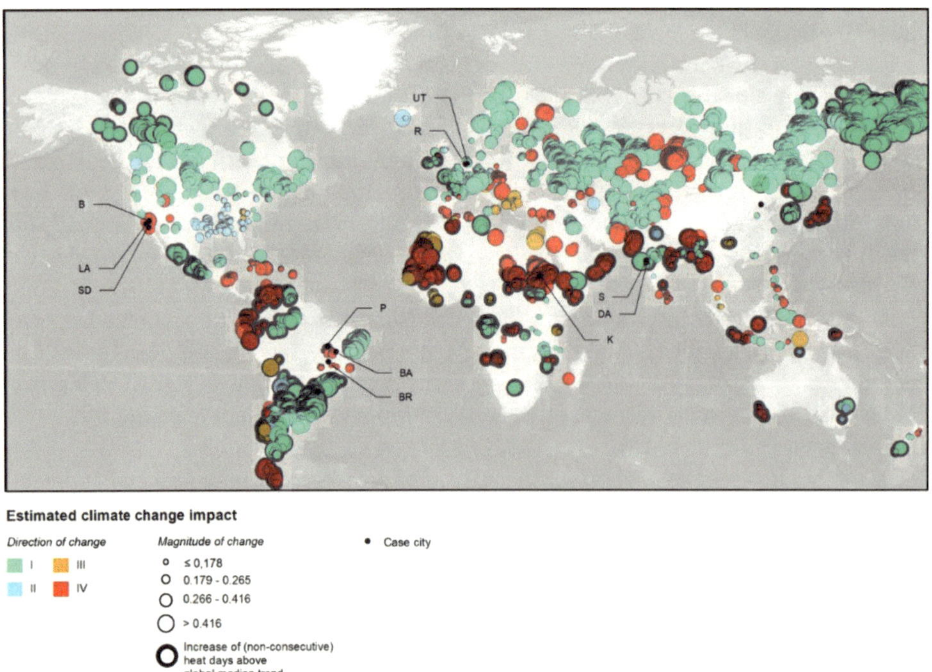

Figure 3.
Mapping direction of temperature change from 1901 until 2014 and rainfall from 1901 until 2013 in cities [52].

Decision-makers of cities need improved information, that regularly updated, about human behavior and perceptions and how they relate to climate change, both globally and locally. Considering human behavior in cities and linking it to downscaled climate projections and remotely sensed observations of urban form, land cover, land-use patterns and social-demographic information from national and international databases, has the potential for improved decision-making to drive a much more nuanced and highly spatially resolved platform. Over the past decade, with the advancement of the digital social sciences and big data, as well as the increasing use of social media data (SMD) in geographical studies, new opportunities have emerged for augmenting urban systems and climate impacts research and expanding them [57].

Geocoded SMD, which comes from social media users (e.g., Twitter, Facebook, Instagram) opens up a significant new opportunity to fill data gaps and address many of the obstacles that prevent researchers and practitioners from understanding the human behavior component of urban system dynamics and climate change.

SMD and other big data let researchers to ask a wide range of spatially explicit questions at an unprecedented scale. Most of the time, social media allows users to manually select the location from where they post a message, or automatically adds it via geolocation tracking services. However, at present, geo-located tweets and Flickr photographs represent a small portion of the overall volume of SMD (e.g., only %1 of all tweets are tweets that geocoded via GPS constitute) [58] the sheer quantity of these data makes them worth investigating. Geotagged tweets can reinforce traditional control data (e.g. remotely sensed images, roads, parcels). For instance, for modeling population distribution, geo-located Twitter messages can serve as control data [59].

Research using geo-located SMD is also starting to take shape to study socioeconomic disparities and their relationship to climate impacts in cities. For example, crowd-sourced data from Foursquare users in London, had been used to be a reliable proxy for the localization of income variability and highlighting places which are more at risk across the city [60].

Yet, mapping based on data that are demographically unrepresentative, can also regenerate spatial segregation and give an unfair picture of the places which matter citywide [61]. The same is true for global-scale analyses. The amount of geocoded tweets widely varies among nations. The United States and Brazil are some of the countries whit the highest proportion of geocoded and non-geocoded tweets, while countries such as Norway and Denmark record considerably lower values [62].

The emergence of various types of big data provides interesting options for evaluating how people use and respond to urban events, policies, programs and designs to adapt and mitigate climate change. New types of data can be an important source for examining the use, value and social equity of specific spaces in the city, that provide refuge during climate driven extreme events, such as parks, vacant areas and open spaces that can provide, like cooling during heat waves. Working with big data can provide opportunities for multi-year to decadal datasets to understand the interactions between human and nature in the city and could be crucial to evaluate progress on examining influences of climate change and of mitigation options in cities [57].

Various sources of big data have already been helpful for awareness of disaster risk management and climate adaptation planning. In [63], authors for assessing the desired location and capacity of flood evacuation shelters, used volunteered geographic information through SMD as a source; while, in [64], researchers used SMD sourced observations of flooding to develop a method for estimating flood extent in Jakarta. In addition, following the devastating impact of Hurricane Sandy in New

York City, scientists used SMD in Twitter to reveal the geographies of a range of social processes and actions that happened shortly after the event [65].

In another research, Twitter data collected have been used during the devastating Sendai earthquake in Japan to assess social networks and build a database to study the human landscape of post-disaster effects [66]. Understanding interactions between climate change and fire prone landscapes is another major concern for adapting with climate change and for reducing the disaster risk. In [67], authors were able to use SMD to evaluate spatial patterns of situational awareness during the Horsethief Canyon Fire in Wyoming, besides they demonstrated the usefulness of SMD for actionable content during a crisis.

Another promising route which has been used in previous research, is the combination of satellite data with other datasets and analyzing it by ML. For example, in [68], authors proposed an accurate, scalable and cheap method for estimating consumption expenditure and asset wealth from high-resolution satellite imagery. In this research, using satellite data from five African countries: Nigeria, Tanzania, Uganda, Malawi, and Rwanda, they showed how a CNNN can be trained to detect image features that can explain up to 75% of the variation in local-level economic outcomes, which result in estimation of poverty levels via satellite data.

Another point is that, big data can become a central tool for online monitoring of urban risks and climate policies, made possible by sensor-based cities and the large amounts of data typically generated by their residents through social media. Applications include [69]:

- Use real-time data extracted from local weather stations, rainfall and sewer gauges to collect real-time data in hydrodynamic models for improved flood prediction;

- Combine local high frequency observations, with regional monitoring and forecasts, along with tracking of geospatial social messaging (e.g., posts about occurring events) to provide improved early warning about potential effects;

- Use image processed CCTV feeds to understand the risks, for instance, water surface locations and social media feeds to validate in real-time the emergent the flooding patterns;

- Integrate spatially heterogeneous sensor data from flows and movements with geospatial social messaging, CCTV and other data to reach a better understanding of the temporal dynamics of impacts;

- Combine CCTV monitoring with social media data feeds to improve understandings of citizen reaction and response to emerging impacts for optimized hazard mitigation and planning in future;

- Apply knowledge from previous events, like modeling result sets of both risks and impacts, to improve pre-response event from the site to the city-scale for future events.

ML techniques, especially neural networks, are powerful tools for multi-dimensional and complex big data analysis, where complexity needs to be reduced to understand its main drivers [70]. CNNs work well to classify images [71], and are widely used to evaluate land-use patterns [72]. Some researchers have gone even further and combined this approach with the socio-economic data analysis [68, 73].

As mentioned before, authors' work in [68] is very instructive. Using a combination of CNN, daytime satellite imagery and nightlight data, they predict poverty in five African countries at a village scale. For this purpose, they did their analysis in three steps. In the first step the CNN is trained on ImageNet [74] to learn how to recognize visual attributes like edges and corners. Second, it was well tuned so as to be able to predict intensities of the night-time in daytime images. Nightlights are a universally consistent poverty predictor. Therefore, the model was trained to focus on the aspects in daytime imagery, which are relevant to poverty estimation. In the third and final step, socio-economic survey data was used to train ridge-regression models on both household surveys and the image features from the previous two steps. Their approach used night-time data as a globally consistent, but very noisy proxy for poverty in an intermediate step and eventually explains %37–55 of average household consumption, and %55–75 of the variation in average household asset wealth. While it used publicly available data, it delivered better results than cellphone-based studies and outperforms products that rely solely on nightlights.

Recently another study used data extracted from Google Street View images and ML methods, such as v-support regression and feature extraction, to estimate high income areas in US cities [38]. From another perspective, phone records were used to reveal detailed mobility patterns for improving the understanding of travel behavior and traffic management [75].

3.2 High-precision measurement

An example of the activity recognition task is transportation mode detection can be in which data from smartphone sensors carried by users are employed to deduce what transportation mode the individuals have used. Microelectromechanical systems (MEMS), such as gyroscopes and accelerometers are embedded in most smartphone devices [76] from which the data can be obtained at high frequencies.

Nowadays, smartphones have powerful sensors like Global Positioning System (GPS), accelerometer, light sensors, gyroscope, etc. Having such sensors that embedded in a small device carried in all life activities has allowed researchers to investigate new research areas. Some of the benefits of these smart devices include ability to send and receive data through various ways (e.g. Wi-Fi/cellular network/Bluetooth), ubiquity and processing data [77]. The knowledge of individuals' transport mode can be adopted in many applications and also can facilitate several tasks, as follows:

1. Knowing transportation mode is a necessary part of urban planning for transportation, which is usually examined through questionnaires or telephone interviews or travel diaries [77, 78]. Most of the time, this traditional method of polling is inaccurate, expensive, limited to a specific area, and not up-to-date [79].

2. As environmental applications, by obtaining the transport mode, the carbon footprint and the amount of calories burnt of individuals can be determined. Besides, health situation and physical activities can be monitored, the risk exposure can be tracked, and the environmental influences of one's activities can be examined [80].

3. Other applications such as giving real-time information to users with the knowledge of speed and transport mode from the them as probes [77, 81], offering individuals with personalized messages and advertisements based on their transportation mode [77].

Several studies have used GPS data for classification purposes. However, using GPS sensors have some limitations, such as: in shielded areas like tunnels, GPS information is not available and the GPS signals may be lost especially in high dense locations, which results in erroneous position information. In addition, the GPS sensor consumes power a lot, so sometimes users turn it off to save the battery [79, 81]. Some research focus on developing detection models using ML techniques and data obtained from smartphone sensors like gyroscope, accelerometer and rotation vector, without GPS data [82]. In this way, it has the advantage of considering multiple sensors even without using GPS, the transportation modes can be identified.

4. Evaluation methods

Given the area of research, there are always some standard methods to evaluate a system that uses a ML algorithm. In addition, there should be some standard datasets the are prepared for the learning process (training, tuning and testing). Therefore, in the following the evaluation metrics and datasets in GIS for urbanization are introduced.

4.1 Metrics

There is no single connotation for the word "quality", because it is difficult to define quality with an absolute concept. Obviously, the data quality within software systems relates to the benefits that can be achieved by an organization. Furthermore, it is dependent on various aspects. Thus, to measure data quality accurately, one unique feature has to be chosen for considering the contribution of other attributes of the data quality as a whole. Following aspects can be used to describe the data quality (**Table 2**) [83].

It should be noted that by getting a high score in any of the mentioned dimensions, does not simply mean that a high quality data has been achieved. For example, the timelines may only matter in terms of correctness (correct user information is available, but if it is not updated, then it is useless). Sometimes, these features complement to each other [83].

Dimensions	Definition
Relevance	The importance of each piece of information stored in the database.
Reliability	The sources of data are reliable.
Correctness	The real world situation is represented by each set of stored data.
Timeliness	The data has been updated on time and with adequate frequency
Precision	The accuracy of the stored data is enough to characterize it.
Unambiguous	Each piece of data carries a unique meaning.
Accuracy	The level of data that can be accurately represented.
Objectivity	Data is objective: do not need people to judge, interpret, or evaluate.
Security	Access is secure and limited.
Completeness	The absence of the essential data: how much available data is missing.

Table 2.
Patterns of data quality dimensions.

The goals of data quality metrics are multi-dimensional. Indeed, they can set information quality objectives for data creators and managers to achieve, set standards for data to be produced, acquired and curated, and introduce measurement methods for quality judgment.

These metrics include rules that determine the thresholds of meeting appropriate professional expectations and govern the measurement of data quality aspects and levels. In order to configure and organize the rules, a basic structure is needed to distinguish the transformation process from data quality expectations to a set of applicable claims and to prevent unprofessional conduct [84].

Defining dimensions of data quality metrics can meet some purposes. Most of the time, the dimensions are classified according to accepted standard of scholarly activities within an academic discipline as well as other related disciplines that use the data. Scientists have developed several sets of data quality dimensions [85].

The dimension categories differ from each other, according to the academic field(s) in which data are regulated or by the different researchers' understanding and preference. Not only their dimensions are categorized differently among scholars, but also their definitions vary according mostly to different types of data. In practice, variations exist, such as integrity may be described in a different way to measurement adjusted strategies, and accuracy may be calculated at different levels of explanation [85].

Landslide susceptibility mapping (LSM) is a prime step in implementing the immediate disaster management planning and risk mitigation measures. All susceptibility models must be verified for their predictions accuracy. An unverified prediction model and susceptibility maps are nonetheless meaningless and hence do not have any scientific significance. The issue of LSM validation have tackled by many studies [86].

Several LSM approaches have been developed and described in numerous papers. These approaches are mainly divided into three groups: heuristic, deterministic and statistical techniques.

The heuristic techniques are based on the expert's knowledge to group landslide-prone areas into several ranks from high to low classes. It is often used for susceptibility mapping in large areas. While, deterministic techniques rely on numerical modeling of the physical mechanism that controls slope failure. However, they are not suitable for a large-scale mapping, due to their problematic and impractical need for a huge array of data, namely rock mechanical properties, the wetness and soil saturation and soil depth. Statistical and probabilistic techniques including bivariate, multivariate statistical methods, certainty factor, as well as knowledge-based techniques such as ANNs and fuzzy logic approaches are promising methods for predicting the landslides [87].

In most cases, the models are tested with an independent set of data, which was not used for training the model. In [88], authors reported a three following approach to obtain an independent sample of the landslide for validation purpose [87].

1. From the entire landslide inventory map of the study area, two sets of randomly divided landslide polygons should be created, one for the susceptibility analysis and one for validation the models;

2. In a part of the whole study area, the susceptibility analysis should be performed; the obtained result should be tested in another part, distinctly with different landslides;

3. The analysis should be performed using landslides happened in specific period and validation should be carried using landslides occurred in a different

period. This is the most sufficient to test the validity of the prediction mode, however, the toughest to apply as it requires knowledge of the temporal distribution of landslides over an adequately long-time spans.

As an example, image classification is not valid without evaluating its accuracy. The source of errors could stem from the classification itself, image recording, inappropriate training data and so on, however, in accuracy evaluation it is assumed that all differences between classification results and reference data comes from the classification errors.

Confusion matrix is one of the most common methods that evaluates classification accuracy. This matrix contains a category comparison of relationship between known, ground-truth data and classification results for the same category.

The overall accuracy of the classification process, is measured in percent and indicates the number of pixels which correctly classified divided by the total number of pixels. Kappa coefficient is a measure of overall statistical agreement. It measures the overall agreement of classification results, excluding agreement acquired, not on purpose, but by chance [89].

4.2 Data

From the very first satellite launched in 1972 till the Landsat 8, launched in 2013, Landsat satellite data have been recognized as a source of objective and reliable information. These missions provide high quality worldwide multispectral data and have been successfully used in countless applications in science [90].

The Landsat archive has provided multispectral data over the Earth for about 40 years. This fact makes Landsat data an attractive information source for studies related to change detection, especially for identifying land use and land cover changes indications.

World population was more than 7 billion at the time of the latest Landsat, Landsat 8. Considering the valuable information about changes to Earth's land surface for more than 40 years, the Landsat program has given decision makers a reliable source for managing Earth's resources for the planet's burgeoning population with integral information about the World's food, water, forests and how land resources are being used [90].

Imagery from these satellites is distributed for free and was obtained from the USGS website: http://earthexplorer.usgs.gov/.

Landsat 5 had Multi-Spectral Scanner (MSS) and Thematic Mapper (TM) sensors. TM sensor has 6 spectral bands with the resolution – 30 m and 1 thermal infrared band with resolution of 120 m (**Table 3**) [91].

Landsat 7 has Enhanced Thematic Mapper Plus (ETM+) sensor with 6 multispectral bands with 30 m resolution, 1 thermal band with the resolution of 60 m and 1 panchromatic band with 15 m resolution (**Table 4**). Bands 1–5&7 were used for LULC classification, while band 6 for LST extraction in both cases of Landsat 5 and 7 [91].

High-resolution maps of settlements and urban footprints form the basis for an integrated evaluation of global settlement patterns. In the past decade, there has been rapid progress in preparation of such maps. New satellite technology and improved data processing using ML have facilitated rapid improvement in their accuracy and resolution. The MODIS 500 urban land cover [92], until recently represented the state of the art in urban land cover datasets [93]. It is now outperformed by both the Global Urban Footprint (GUF) dataset which have higher resolution and accuracy than any other urban land cover dataset [94], even if comparing it to the high quality Global Human Settlement Layer (GHSL) [95, 96] or GlobeLand 30 [97]. The GUF attributes a binary urban footprint at a resolution as

Band	Spectral band	Resolution
1	0.45–0.52 μm	30 m × 30 m
2	0.52–0.60 μm	30 m × 30 m
3	0.63–0.69 μm	30 m × 30 m
4	0.76–0.90 μm	30 m × 30 m
5	1.55–1.75 μm	30 m × 30 m
6	10.4–12.5 μm	120 × 120 m
7	2.08–2.35 μm	30 m × 30 m

Table 3.
Landsat 5 TM bands.

Band	Spectral band	Resolution
1	0.45–0.515 μm	30 m × 30 m
2	0.525–0.605	30 m × 30 m
3	0.63–0.69 μm	30 m × 30 m
4	0.75–0.90 μm	30 m × 30 m
5	1.55–1.75 μm	30 m × 30 m
6	10.4–12.5 μm	60 m × 60 m
7	2.09–2.35 μm	30 m × 30 m

Table 4.
Landsat 7 ETM+ bands.

high as 0.4″ (approximately 12 m) at the equator and 0.6″ in the mid-latitudes on a global coverage. Also it is freely available for scientific use. This high resolution constitutes a paradigm shift in studying urban extent for cities around the world.

The importance of the satellite imageries for evaluating urbanization by measuring land use and land cover change for cities and their surroundings, is undeniable. Remote sensing (RS) is a reliable data source, which provides spatially consistent coverage of large areas with temporal frequency and high spatial detail. Besides, it is useful for analyzing phenomenon that is time dependent, such as urban expansion [98]. Therefore, RS is an accurate and effective data source for monitoring expansion of metropolitans, especially in cases that information related to the land use management is inconsistent and inappropriate.

This is a list of some other datasets that provide information related to GIS for urbanization:

1. GLOBAL Map[1]: It is a set of digital maps that cover the entire world to express the status of global environment, accurately. It is developed through the cooperation of National Geospatial Information Authorities (NGIAs) in the world. The Global Map provides eight main map themes at a nominal ground resolution of 1 km for raster data and at a scale of 1:1,000,000 for vector data. These themes are:

    - Transportation
    - Boundary

[1] https://nationalmap.gov/small_scale/atlas-ftp-global-map.html?openChapters=chptrans#

- Drainage
- Population Centers
- Elevation
- Vegetation
- Land Cover
- Land Use

2. Gridded Population of the World (GPW)[2]: It is the dataset of NASA's socioeconomic data and applications center, which includes raw population, and population density of the past, current and future prediction. The purpose of GPW is to provide a spatially disaggregated population layer that is compatible with datasets from social, economic, and Earth science disciplines, and RS. This data is globally consistent and spatially explicit for research, decision-making and communication.

3. World Bank Geodata[3]: In this data, a wide range of World Bank datasets converted to KML format, including GNP, schooling and financial data.

4. Global ADMINISTRATIVE Areas[4]: Administrative areas in this database are countries and lower level subdivisions such as provinces and departments. The latest version is 3.6 and it was released in 2018. It restricts 386,735 administrative areas, and scientists can download the spatial data by country.

5. Armed CONFLICT Location and Event Dataset[5]: This data includes all reported conflict events in 50 countries in developing world, from 1997 to present.

6. Global Rural-Urban Mapping Project (GRUMP)[6]: It is the dataset from NASA'S socioeconomic data and applications center, which includes information on rural and urban population balances.

7. Open Street Map (OSM)[7]: Crowdsourced data for the whole world, which contains many important things like points of interest, buildings, roads and road names, ferry routes, etc.

8. Geohive[8]: the initiative is made available by Ordnance Survey Ireland for easy access to public spatial data, and includes population and county statistics. it is not provided in GIS data formats, but it is easily convertible from CSV.

---

[2] https://sedac.ciesin.columbia.edu/data/collection/gpw-v4

[3] https://sourceforge.net/projects/googleworldbank/

[4] https://gadm.org/

[5] https://www.acleddata.com/

[6] https://sedac.ciesin.columbia.edu/data/collection/grump-v1

[7] https://learnosm.org/en/osm-data/osm-in-qgis/

[8] https://geohive.ie/

4.2.1 Urban landscapes

The World Bank, in south and east Asia, has explored the patterns, consequences and policy implications related to spatial development of cities by outlining the increasing availability of spatial data and developments in analytics. Data from Earth observation (EO) satellite can give valuable results which are useful for measuring urban growth over a wide range of spatial and temporal scales, especially when combined with data from other sources. The resulting digital urban maps give an accurate, up-to-date and cost effective resource to assist governments in understanding the nature of urban development and making informed decisions. EO datasets allow for harmonized and standardized measurements. Also, they enable planners to make spatially and temporally consistent comparisons and global assessment. In addition, they are particularly significant for monitoring and understanding the evolution of cities. For instance, allowing authorities to know when built-up areas spill across formal administrative boundaries. This shows the need to cooperate with adjoining administrative areas on issues like collecting garbage or connective infrastructure [99].

The World Bank has created a database to analyze the speed, magnitude and spatial form of urbanization in EO data. These data help researchers examine the drivers and influences of the urbanization nature and how the urban landscape has evolved into its current state. It offers a basis for understanding the effects of policy change and identifying priorities for new initiatives. In particular, the focus is on exploring the institutional frameworks for urban management, like mechanisms to coordinate service delivery across administrative jurisdictions, investment for example in transport and other network infrastructure and regulation such as zoning and pricing of services.

About twelve years ago, the World Bank launched the "Earth Observation for Development" initiative. So, data in areas where data are commonly scarce and unreliable, are provided. Such information is useful for building project fundamentals against which progress can be gauged, high priority issues identified and mitigation measures determined. Focus of this project is on areas like metropolitan development and related fields including disaster risk management, the environment, water and energy. The bank has also developed the Urban Management and Analysis (PUMA) platform to facilitate more collaboration between policymakers and other development stakeholders, toward these purposes. By using this tool, users with no GIS experience would be able to access, analyze and share urban spatial data in an interactive and customizable way [100].

These activities have resulted in more than 30 technical helping projects that done for urban planners and partners, in the period 2008–2018. As a result, highly specialized big data mapping products and monitoring systems that leverage EO data for South Asian cities have been launched.

4.2.2 Megacities

In the South Asia Megacities Improvement Program, EO big data was used to analyze 20 years of urban expansion in the metropolitan areas of Delhi, Mumbai and Dhaka. These data make it possible to measure the qualitative and quantitative aspects of transformation, like the distribution and density of urban sprawl, the growth rate of built-up areas and urban land use change. This information helps analysts to trace how informal settlements grow outside the cities' boundaries, and to understand the drivers of land use [99]. Therefore, some important insights into land cover and use in the three cities revealed (**Figure 4**). Furthermore, it showed

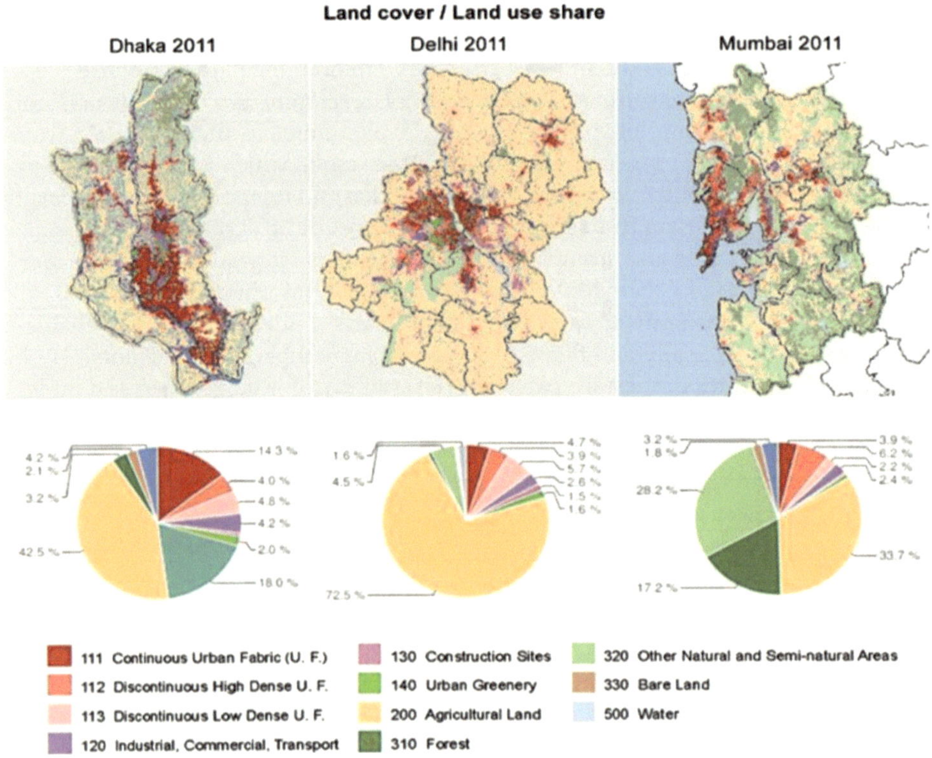

Figure 4.
Sample visualizations from the South Asia geospatial analysis [101].

the percentage of land taken by settlements and industrial build-up, agriculture, natural or semi-natural vegetation and forest [101].

By using it, urban planners and development stakeholders could understand existing demands and plan for future needs. For instance, in Delhi, the maps illustrate that with industrial development, the urban expansion is accelerated. This mostly happened between 2003 to 2010; however, a considerable increase in construction sites shows that it will continue in the future, so it must be planned [101].

By using digitized spatial data, analysts would be able to study the target at different administrative levels: metropolitan, city, district or sub-district, and also other non-administrative units. These datasets make it possible to aggregate flexibly. One example is showing the proportion of sprawl by district, its density, the drivers of urban change and class evolution within urban areas. Together with environmental or socio-economic data, the data can prepare information on the proportion of population to urban growth, and can measure indicators like compactness, the ratio of green space to citizens, and the accessibility of these areas.

The results of applying EO big data can be crucial for coordination between public, private and household investment in infrastructure, productive capital and housing, respectively. Thus, policymakers would be able to promote optimal spatial and transportation links between businesses, affordable housing and commercial units, health and education services and recreational areas. In addition, these views can be applied to support rural-to-urban migrants and ensuring that rapid urbanization is inclusive. Since EO big data methods spread across the world's megacities, and are refined and adapted, they will provide valuable tools to policymakers, and greater benefits for the citizens of the future.

4.2.3 Residential cities

EO big data approaches are also contribute to drive sustainable urban development. The mentioned research on the use of high-resolution satellite data for poverty mapping, draws emerging techniques that can show fast changing urban areas in near real-time. These methods can determine built-up area, density of cars and buildings, and types of roofing and road. Via ML techniques and image processing algorithms, also they can calculate whether buildings are more rectangular or have more chaotic angles, that indicates higher poverty level, and construct poverty indicators like the ratio of paved roads in an area. So, stakeholders can target their interventions exactly where they are mostly useful [101].

All in all, analysis of EO big data can be an important tool for managing city development in low-income countries. It can measure and track the urban expansion and highlight the drivers of economic growth. This result in better understanding the factors contributing to inefficiencies and inequality in urban areas, and providing optimized policies. Besides, they can create flexibility in urban environments, so that residents, businesses and systems can adapt to persistent stresses or shocks. Also they can provide residential cities that meet their residents' needs.

5. Conclusions

This book chapter briefly introduced ML and past research about the application of ML algorithms for processing of daily satellite imagery. It has been demonstrated several aspects of detecting and classification of Earth features merging into local geographical and geodetical system with further GIS development. The main purpose of the chapter is to provide existing resources for researchers to be aware of the up-to-date status of development of ML application in GIS in particular in studies of megacities.

The real potential of ML in GIS is not sufficiently developed yet. On one hand, both fields intersect in analytical discussions. At the same time, most GIS applications which are desirable for ML implementation, are driven by conventional approach and standard tools of commercial GIS packages.

Merging GIS and ML offers a potential mechanism to reduce the cost of analysis of spatial information by decreasing the amount of time spent on data interpretation. This integration allows the interpretive outcome from a small area to be transferred to a larger, geographically similar area, without the extra time and expense of putting geographers in the field for a time sufficient to cover geographical area.

ML can be considered both as a science and as engineering, depending on the goal. This technology is often seen as part of computing; however, it has links with various other areas including philosophy, psychology and linguistics. Its techniques can provide benefits within GIS over traditional methods, like statistical analysis, especially if data show some form of non-linearity. Thanks to such an opportunity of ML/GIS technology makes most successfully to apply for monitoring and observation consequences of megacity development.

Most people are unaware that they use artificial intelligence in their daily life. Finding solutions to decision-making issues by using models that allow decision makers to express their limitations and imprecise concepts that are used with large volume of geographic data, costs a lot. This chapter is expected to open opportunity to understand clearly fundamental aspects of ML/GIS development with basically related to the megacity studies.

Author details

Nasim Tohidi and Rustam B. Rustamov*
Alzahra University, Tehran, Iran

*Address all correspondence to: r_rustamov@hotmail.com

References

[1] Ray, S. A Comparative Analysis and Testing of Supervised Machine Learning Algorithms. 2018.

[2] Ayodele, T. Types of Machine Learning Algorithms. In: New Advances in Machine Learning. InTech; 2010. DOI: 10.5772/9385

[3] 15th Annual Demographia World Urban Areas. Demographia. Parliament of Hungary, Budapest. 2019.

[4] Fenta, A.A. Yasuda, H. Haregeweyn, N. Sewale Belay, A. Hadush, Z. Gebremedhin, M.A. Mekonnen, G. The dynamics of urban expansion and land use/land cover changes using remote sensing and spatial metrics: the case of Mekelle City of northern Ethiopia. International Journal of Remote Sensing. 2017:38:4107–4129. DOI: 10.1080/01431161.2017.1317936

[5] Akintunde, J.A. Adzandeh, E.A. Fabiyi, O.O. Spatio-temporal pattern of urban growth in Jos Metropolis, Nigeria. Remote Sensing Applications: Society and Environment. 2016:4:44–54. DOI: 10.1016/j.rsase.2016.04.003

[6] Farooq, M. Muslim, M. Dynamics and forecasting of population growth and urban expansion in Srinagar City – A Geospatial Approach. In: The International Archives of the Photogrammetry, Remote Sensing and Spatial Information Sciences, Hyderabad; 2014. DOI: 10.5194/isprsarchives-XL-8-709-2014

[7] Dereli, M.A. Uğur, M.A. Polat, N. Yalçın, M. Spatio-Temporal Analysis of Urban Expansion Using Remote Sensing Data. In: 17th International Multidisciplinary Scientific GeoConference SGEM2017; Albena, Bulgaria; 2017. DOI: 10.13140/RG.2.2.35243.34084/1

[8] Kivilcim, C.Ö. Duran, Z. Impact of rapid urbanization on the morphology of historical Istanbul: Uskudar district case study. Fresenius Environmental Bulletin. 2015:24:14–18.

[9] Wilson, A.G. Urban and regional models in geography and planning. John Wiley and Sons. 1974.

[10] Karakus, C.B. Kavak, K.S. Cerit, O. Determination of variations in land cover and land use by remote sensing and geographic information systems around the city of Sivas (Turkey). Fresenius Environmental Bulletin. 2014:23:667–667.

[11] Son, N.T. Chen, C.F. Chen, C.R. Thanh, B.X. Vuong, T.H. Assessment of urbanization and urban heat islands in Ho Chi Minh City, Vietnam using Landsat data. Sustainable Cities and Society. 2017:30:150–161. DOI: 10.1016/j.scs.2017.01.009

[12] Herold, M. Goldstein, N.C. Clarke, K.C. The spatiotemporal form of urban growth: measurement, analysis and modeling. Remote Sensing of Environment. 2003:86:286–302. DOI: 10.1016/S0034-4257(03)00075-0

[13] LStefanov, W. Ramsey, M.S. Christensen, P.R. Monitoring urban land cover change: An expert system approach to land cover classification of semiarid to arid urban centers. Remote Sensing of Environment. 2001:77:173–185. DOI: 10.1016/S0034-4257(01)00204-8

[14] Hoffhine Wilson, E. Hurd, J.D. Civco, D.L. Prisloe, M.P. Arnold, C. Development of a geospatial model to quantify, describe and map urban growth. Remote Sensing of Environment. 2003:86:275–285. DOI: 10.1016/S0034-4257(03)00074-9

[15] Weng, Q. Land use change analysis in the Zhujiang Delta of China using satellite remote sensing, GIS and

stochastic modelling. Journal of Environmental Management. 2002:64: 273–284. DOI: 10.1006/jema.2001.0509

[16] Chrysoulakis, N. Mitraka, Z. Stathopoulou, M.I. Cartalis, C. A comparative analysis of the urban web of the greater athens agglomeration for the last 20-years period on the basis of landsat imagery. Fresenius Environmental Bulletin. 2013:22:2139–2143.

[17] Sariyilmaz, F.B. Musaoglu, N. Tanik, A. Investigation of land use/cover changes of Sazlidere basin by using band ratio for builtup area (BRBA). Fresenius Environmental Bulletin. 2017:26:39–45.

[18] Skapura, D.M. Building Neural Networks, New York: Addison-Wesley Professional; 1996.

[19] Tayyebi, A. Delavar, M.R. Saeedi, S. Amini, J. Alinia, H.S. Monitoring land use change by multi-temporal Landsat remote sensing imagery. In: The International Archives of the Photogrammetry, Remote Sensing and Spatial Information Sciences. Beijing. 2008. DOI: 10.13140/2.1.2736.3204

[20] Tayyebi, A. Pijanowski, B.C. Tayyebi, A.H. An urban growth boundary model using neural networks, GIS and radial parameterization: An application to Tehran, Iran. Landscape and Urban Planning. 2011:100:35–44. DOI: 10.1016/j.landurbplan.2010.10.007

[21] Li, J. Huang, X. Gong, J. Deep neural network for remote-sensing image interpretation: status. National Science Review, pp. 1–4, 2019. DOI: 10.1093/nsr/nwz058

[22] Liu, X. Han, F. Ghazali, K.H. Mohamed, I.I. Zhao, Y. A review of Convolutional Neural Networks in Remote Sensing Image. In: Proceedings of the 2019 8th International Conference on Software and Computer Applications. 2019. DOI: 10.1145/3316615.3316712

[23] Tohidi, N. Hasheminejad, S.M.H. MOQAS: Multi-objective question answering system. Journal of Intelligent and Fuzzy Systems. 2019:36:3495–3512. DOI: 10.3233/JIFS-181364

[24] Davoodi, M. Mesgari, S.M. GIS-based Route Finding using Ant Colony Optimization and Urban Traffic Data from Different Sources. In: International Conference on Sensors & Models in Remote Sensing & Photogrammetry; Kish Island, Iran; 2015.

[25] Zhang, Y. Zhang, G. Zhao, H. Cao, Y. Liu, Q. Shen, Z. Li, A. A Convenient Tool for District Heating Route Optimization Based on Parallel Ant Colony System Algorithm and 3D WebGIS. International Journal of Geo-Information. 2019:8:1–18. DOI: 10.3390/ijgi8050225

[26] Eldrandaly, K.A. Hassan, M.M. AbdelAziz, N.M. A Modified Artificial Bee Colony Algorithm for Solving Least-Cost Path Problem in Raster GIS. International Journal of Applied Mathematics. 2015:9:147–154.

[27] Davoodi, M. Malekpour Golsefidi, M. Mesgari, S.M. A hybrid optimization method for vehicle routing problem using artificial bee colony and genetic algorithm. In: Joint Conferences of SMPR and GI Research; Karaj, Iran; 2019.

[28] García-Alonso, C.R. Pérez-Naranjo, L.M. Fernández-Caballero, J.C. Multiobjective evolutionary algorithms to identify highly autocorrelated areas: the case of spatial distribution in financially compromised farms. Annals of Operations Research. 2014:219:187–202. DOI: 10.1007/s10479-011-0841-3

[29] Anguelov, D. Dulong, C. Filip, D. Frueh, C. Lafon, S. Lyon, R. Ogale, A. Vincent, L. Weaver, J. Google Street View: Capturing the World at Street Level. Computer. 2010:43:32–38. DOI: 10.1109/MC.2010.170

[30] Goodchild, M. Citizens as Sensors: The World of Volunteered Geography. GeoJournal. 2007:69:211–221. DOI: 10.1007/s10708-007-9111-y

[31] Hays, J. Efros, A.A. Large-Scale Image Geolocalization. Multimodal Location Estimation of Videos and Images. 2015:41–62. DOI: 10.1007/978-3-319-09861-6_3

[32] Liu, L. Zhou, B. Zhao, J. Ryan, B.D. C-IMAGE: city cognitive mapping through geo-tagged photos. GeoJournal. 2016:81:817–861. DOI: 10.1007/s10708-016-9739-6

[33] Zhang, F. Zhang, D. Liua, Y. Lin, H. Representing place locales using scene elements. Computers. Environment and Urban Systems. 2018:71:153–164. DOI: 10.1016/j.compenvurbsys.2018.05.005

[34] Salesses, P. Schechtner, K. Hidalgo, C. The Collaborative Image of The City: Mapping the Inequality of Urban Perception. PLoS ONE. 2013:8. DOI: 10.1371/journal.pone.0068400

[35] Arase, Y. Xie, X. Hara, T. Nishio, S. Mining people's trips from large scale geo-tagged photos. In: Proceedings of the 18th ACM international conference on Multimedia; Firenze, Italy; 2010.

[36] Crandall, D. Backstrom, L. Huttenlocher, D.P. Kleinberg, J.M. Mapping the world's photos. In: Proceedings of the 18th International Conference on World Wide Web; Madrid, Spain; 2009.

[37] Less, E.L. McKee, P. Toomey, T. Nelson, T. Erickson, D. Xiong, S. Jones-Webb, R. Matching study areas using Google Street View: A new application for an emerging technology. Evaluation and Program Planning. 2015:53:72–79. DOI: 10.1016/j.evalprogplan.2015.08.002

[38] Glaeser, E.L. Kominers, S.D. Luca, M. Naik, N. Big Data and Big Cities: The Promises and Limitations of Improved Measures of Urban Life. Economic Inquiry. 2018:56:114–137.

[39] Dubey, A. Naik, N. Parikh, D. Raskar, R. Hidalgo, C.A. Deep Learning the City: Quantifying Urban Perception At A Global Scale. in European Conference on Computer Vision (ECCV), 2016. DOI: 10.1007/978-3-319-46448-0_12

[40] Naik, N. Philipoom, J. Raskar, R. Hidalgo, C. Streetscore – Predicting the Perceived Safety of One Million Streetscapes. In: IEEE Conference on Computer Vision and Pattern Recognition Workshops, 2014.

[41] Ordonez, V. Berg, T.L. Learning High-Level Judgments of Urban Perception. In: European Conference on Computer Vision, 2014. DOI: 10.1007/978-3-319-10599-4_32

[42] Porzi, L. Rota Bulò, S. Lepri, B. Ricci, E. Predicting and Understanding Urban Perception with Convolutional Neural Networks. In: Proceedings of the 23rd ACM international conference on Multimedia, 2015. DOI: 10.1145/2733373.2806273

[43] He, K. Zhang, X. Ren, S. Sun, J. Deep Residual Learning for Image Recognition. In: 2016 IEEE Conference on Computer Vision and Pattern Recognition (CVPR); Las Vegas; 2016. DOI: 10.1109/CVPR.2016.90

[44] Russakovsky, O. Deng, J. Su, H. Krause, J. Satheesh, S. Ma, S. Huang, Z. Karpathy, A. Khosla, A. Bernstein, M. Berg, A.C. Fei-Fei, L. ImageNet Large Scale Visual Recognition Challenge. International Journal of Computer Vision. 2015:115:211–252. DOI: 10.1007/s11263-015-0816-y

[45] Zhang, S. Zhang, S. Huang, T. Gao, W. Speech Emotion Recognition Using Deep Convolutional Neural Network and Discriminant Temporal Pyramid

Matching. IEEE Transactions on Multimedia. 2018:20:1576–1590. DOI: 10.1109/TMM.2017.2766843

[46] Conneau, A. Schwenk, H. Barrault, L. Lecun, Y. Very Deep Convolutional Networks for Natural Language Processing. ArXiv. vol. abs/1606.01781, 2017.

[47] Ren, S. He, K. Girshick, R. Sun, J. Faster R-CNN: Towards Real-Time Object Detection with Region Proposal Networks. IEEE Transactions on Pattern Analysis and Machine Intelligence. 2017: 39:1137–1149. DOI: 10.1109/TPAMI.2016.2577031

[48] Glorot, X. Bengio, Y. Understanding the difficulty of training deep feedforward neural networks. In: Proceedings of the 13th International Conference on Artificial Intelligence and Statistics (AISTATS); Sardinia, Italy; 2010.

[49] Zhang, F. Zhou, B. Liu, L. Liu, Y. Fung, H.H. Lin, H. Ratti, C. Measuring human perceptions of a large-scale urban region using machine learning. Landscape and Urban Planning. 2018: 180:148–160. DOI: 10.1016/j.landurbplan.2018.08.020

[50] Herold, M. Scepan, J. Clarke, K. The Use of Remote Sensing and Landscape Metrics to Describe Structures and Changes in Urban Land Uses. Environment and Planning. 2002:34: 1443–1458. DOI: 10.1068/a3496

[51] McPhearson, T. Pickett, S.T.A. Grimm, N.B. Niemelä, J. Alberti, M. Elmqvist, T. Weber, C. Haase, D. Breuste, J. Qureshi, S. Advancing Urban Ecology toward a Science of Cities. BioScience. 2016:66:1–15. DOI: 10.1093/biosci/biw002

[52] Scheuer, S. Haase, D. Volk, M. Integrative assessment of climate change for fast-growing urban areas: Measurement and recommendations for future research. PLoS ONE. 2017:12. DOI: 10.1371/journal.pone.0189451

[53] Wood, S.A. Guerry, A. Silver, J.M. Lacayo-Emery, M. Using social media to quantify nature-based tourism and recreation. Scientific Reports. 2013:3. DOI: 10.1038/srep02976

[54] Mora, H.M. Pérez-del H.R. Paredes-Pérez, J.F.; Sirvent, R.M. Analysis of Social Networking Service Data for Smart Urban Planning. Sustainability. 2018:10. DOI: 10.3390/su10124732

[55] Martí, P. Serrano-Estrada, L. Nolasco-Cirugeda, A. Social Media data: Challenges, opportunities and limitations in urban studies. Computers, Environment and Urban Systems. 2019: 74:161–174.

[56] Knox, E.G. The Slippery Slope of Material Support Prosecutions: Social Media Support to Terrorists. The Hastings law journal. 2014:66:295–329.

[57] Ilieva, R.T. McPhearson, T. Social-media data for urban sustainability. Nature Sustainability. 2018:1:553–565. DOI: 10.1038/s41893-018-0153-6

[58] Crampton, J.W. Graham, M. Poorthuis, A. Shelton, T. Stephens, M. Wilson, Matthew W. Zook, M.A. Beyond the Geotag: Situating 'Big Data' and Leveraging the Potential of the Geoweb. Cartography and Geographic Information Science. 2013:40:130–139. DOI: 10.1080/15230406.2013.777137

[59] Lin, J. Cromley, R. Evaluating geo-located Twitter data as a control layer for areal interpolation of population. Applied Geography. 2015:58:41–47. DOI: 10.1016/j.apgeog.2015.01.006

[60] Quercia, D. Saez-Trumper, D. Mining Urban Deprivation from Foursquare: Implicit Crowdsourcing of City Land Use. IEEE Pervasive Computing. 2014:13:30–36. DOI: 10.1109/MPRV.2014.31

[61] Cranshaw, J. Schwartz, R. Hong, J. Sadeh, N.M. The livehoods project: Utilizing social media to understand the dynamics of a city. In: Proceedings of the Sixth International Conference on Weblogs and Social Media; Dublin, Ireland; 2012.

[62] Graham, M. Hale, S.A. Gaffney, D. F. Where in the World Are You? Geolocation and Language Identification in Twitter. The Professional Geographer. 2013:66:568–578. DOI: 10.1080/00330124.2014.907699

[63] Kusumo, A.N.L. Reckien, D. Verplanke, J.J. Utilising volunteered geographic information to assess resident's flood evacuation shelters. Case study: Jakarta. Applied Geography. 2017:88:174–185. DOI: 10.1016/j.apgeog.2017.07.002

[64] Brouwer, R. Hadzhiyska, D. Ioakeimidis, C. Ouderdorp, H. The social costs for marine litter along the European coasts. Ocean & Coastal Management. 2017:138:38–49. DOI: 10.1016/j.ocecoaman.2017.01.011

[65] Shelton, A.O. Thorson, J.T. Ward, E. J. Feist, B.E. Spatial semiparametric models improve estimates of species abundance and distribution. Canadian Journal of Fisheries and Aquatic Sciences. 2014:71:1–12. DOI: 10.1139/cjfas-2013-0508

[66] Stefanidis, A. Crooks, A. Radzikowski, J. Harvesting ambient geospatial information from social media feeds. GeoJournal. 2013:78:319–338. DOI: 10.1007/s10708-011-9438-2

[67] Kent, J.D. Capello Jr, H.T. Spatial patterns and demographic indicators of effective social media content during the Horsethief Canyon fire of 2012. Cartography and Geographic Information Science. 2013:40:78–89. DOI: 10.1080/15230406.2013.776727

[68] Jean, N. Burke, M. Xie, M. Davis, W.M. Lobell, D.B. Ermon, S. Combining satellite imagery and machine learning to predict poverty. Science. 2016:353: 790–794. DOI: 10.1126/science.aaf7894

[69] Creutzig, F. Lohrey, S. Bai, X. Baklanov, A. Richard, D. Shobhakar, D. Lamb, W.F. McPhearson, T. Minx, J. Munoz, E. Walsh, B. Upscaling urban data science for global climate solutions. Global Sustainability. 2019:2:1–25. DOI: 10.1017/sus.2018.16

[70] Hinton, G.E. Salakhutdinov, R.R. Reducing the Dimensionality of Data with Neural Networks. Science. 2006: 313:504–507. DOI: 10.1126/science.1127647

[71] Pansombut, T. Wikaisuksakul, S. Khongkraphan, K. Phon-on, A. Convolutional Neural Networks for Recognition of Lymphoblast Cell Images. Computational Intelligence and Neuroscience. 2019:2019:1–12. DOI: 10.1155/2019/7519603

[72] Castelluccio, M. Poggi, G. Sansone, C. Verdoliva, L. Land Use Classification in Remote Sensing Images by Convolutional Neural Networks. arXiv. 2015. DOI: arXiv:1508.00092

[73] Tapiador, F.J. Avelar, S. Corrêa, C.T. Zah, R. Deriving fine-scale socioeconomic information of urban areas using very high-resolution satellite imagery. International Journal of Remote Sensing. 2011:32:6437–6456. DOI: 10.1080/01431161.2010.512928

[74] Deng, J. Dong, W. Socher, R. Li, L.J. Li, K. Li, F.F. ImageNet: a Large-Scale Hierarchical Image Database. In: IEEE Computer Society Conference on Computer Vision and Pattern Recognition (CVPR 2009); Miami, Florida, USA; 2009. DOI: 10.1109/CVPR.2009.5206848

[75] Toole, J.L. Colak, S. Sturt, B. Alexander, L.P. Evsukoff, A. González,

M.C. The path most traveled: Travel demand estimation using big data resources. Transportation Research Part C: Emerging Technologies. 2015:58:162–177. DOI: 10.1016/j.trc.2015.04.022

[76] Susi, M. Renaudin, V. Lachapelle, G. Motion Mode Recognition and Step Detection Algorithms for Mobile Phone Users. Sensors. 2013:13:1539–1562. DOI: 10.3390/s130201539

[77] Stenneth, L. Wolfson, O. Yu, P.S. Xu, B. Transportation Mode Detection using Mobile Phones and GIS Information. In: 19th ACM SIGSPATIAL International Symposium on Advances in Geographic Information Systems; Chicago, IL, USA; 2011.

[78] Xiao, Y. et al. Transportation activity analysis using smartphones. In: IEEE Consumer Communications and Networking Conference (CCNC); Las Vegas; 2012.

[79] Widhalm, P. Nitsche P. Brändie, N. Transport mode detection with realistic Smartphone sensor data. In: Proceedings of the 21st International Conference on Pattern Recognition (ICPR2012); Tsukuba; 2012.

[80] Reddy, S. Mun, M. Burke, J. Estrin, D. Hansen, M. Srivastava, M. Using Mobile Phones to Determine Transportation Modes. ACM Transactions on Sensor Networks. 2010:6. DOI: 10.1145/1689239.1689243

[81] Manzoni, V. Maniloff, D. Kloeckl, K. Ratti, C. Transportation mode identification and real-time CO2 emission estimation using smartphones. Massachusetts Institute of Technology; Cambridge; 2010.

[82] Yazdizadeh, A. Patterson, Z. Farooq, B. An automated approach from GPS traces to complete tripinformation. International Journal of TransportationScience and Technology. 2019:8:82–100. DOI: 10.1016/j.ijtst.2018.08.003

[83] Gaba, S. Gaba, D. Tomar, A. Metric based Approach for Measuring Data Quality. International Journal for Innovative Research in Science & Technology. 2016:2:46–51.

[84] Xia, J. Metrics to Measure Open Geospatial Data Quality. Issues in Science and Technology Librarianship. 2012:68. DOI: 10.5062/F4B85627

[85] Loshin, D. Monitoring Data Quality Performance Using Data Quality Metrics. Informatica Corporation; Redwood City, USA; 2006.

[86] Nohani, E. Moharrami, M. Sharafi, S. Khosravi, K. Pradhan, B. Pham, B.T. Lee, S. Melesse, A.M. Landslide Susceptibility Mapping Using Different GIS-Based Bivariate Models. Water. 2019:11. DOI: 10.3390/w11071402

[87] Dou, J. Yunus, A.P. Bui, D.T. Sahana, M. Chen, C. Zhu, Z. Wang, W. Pham, B.T. Evaluating GIS-Based Multiple Statistical Models and Data Mining for Earthquake and Rainfall-Induced Landslide Susceptibility Using the LiDAR DEM. Remote Sensing. 2019:11. DOI: 10.3390/rs11060638

[88] Irigaray, C. Del Castillo, T.F. El Hamdouni, R. Chacón, J. Evaluation and Validation of Landslide-Susceptibility Maps Obtained by a GIS Matrix Method: Examples from the Betic Cordillera (Southern Spain). Natural Hazards. 2007:41:61–79. DOI: 10.1007/s11069-006-9027-8

[89] Bobrinskaya. M. Remote Sensing for Analysis of Relationships between Land Cover and Land Surface Temperature in Ten Megacities [Thesis]. School of Architecture and the Built Environment Royal Institute of Technology (KTH); Stockholm, Sweden; 2012.

[90] Wulder, M.A. Loveland, T.R. Roy, D.P. Crawford, C.J. Masek, J.G. Woodcock, C.E. Allen, R.G. Anderson,

M.C. Belward, A.S. Cohen, W.B. Dwyer, J. Erb, A. Gao, F. et al. Current status of Landsat program, science, and applications. Remote Sensing of Environment. 2019:225:127–147. DOI: 10.1016/j.rse.2019.02.015

[91] Bobrinskaya, M. Remote Sensing for Analysis of Relationships between Land Cover and Land Surface Temperature in Ten Megacities. Geodesy and Geographic Information Technology; 2012.

[92] Schneider, A. Friedl, M.A. Potere, D.T. Mapping global urban areas using MODIS 500-m data: New methods and datasets based on 'urban ecoregions'. Remote Sensing of Environment. 2010: 114:1733–1746. DOI: 10.1016/j.rse.2010.03.003

[93] Potere, D.T. Schneider, A. Angel, S. Civco, D. Mapping urban areas on a global scale: Which of the eight maps now available is more accurate? International Journal of Remote Sensing. 2009:30:6531–6558. DOI: 10.1080/01431160903121134

[94] Esch, T. Heldens, W. Hirner, A. Keil, M. Marconcini, M. Roth, A. Zeidler, J. Dech, S. Strano, E. Breaking new ground in mapping human settlements from space -The Global Urban Footprint. ISPRS Journal of Photogrammetry and Remote Sensing. 2017:134. DOI: 10.1016/j.isprsjprs.2017.10.012

[95] Pesaresi, M. Ehrlich, D. Freire, S. The Global Human Settlement Layer (GHSL) – New Tools and Geodatasets for Improving Disaster Risk Assessment and Crisis Management. In: 11th International ISCRAM Conference; Pennsylvania, USA; 2014.

[96] Corbane, C. Politis, P. Pesaresi, M. Kemper, T. Siragusa, A. Estimation of Land Use Efficiency from the Global Human Settlement Layer (GHSL). In: QGIS and Applications in Territorial Planning. vol. 3; John Wiley & Sons; 2018. DOI: 10.1002/9781119457121.ch2

[97] Chen, J. Chen, J. GlobeLand30: Operational global land cover mapping and big-data analysis. Science China Earth Science. 2018:61:1533–1534. DOI: 10.1007/s11430-018-9255-3

[98] Jensen, J.R. Cowen, D.C. Remote Sensing of Urban/Suburban Infrastructure and Socio-Economic Attributes. Photogrammetric Engineering & Remote Sensing. 1999:65: 611–622.

[99] Esch, T. Taubenböck, H. Heldens, W. Thiel, M. Wurm, M. Geiss, C. Dech, S. Urban remote sensing - How can earth observation support the sustainable development of urban environments? In: Proceedings, Real CORP; Vienna; 2010.

[100] García, L.E. Rodríguez, D.J. Wijnen, M. Pakulski, I. The World Bank and Remote Sensing. In: Earth Observation for Water Resources Management: Current Use and Future Opportunities for the Water Sector; Washington; World Bank Group; 2016. p. 39–48

[101] Using Geospatial Data to Track Changes in Urbanization [Internet]. Available from: https://olc.worldbank.org/system/files/Using%20Geospatial%20Data%20to%20Track%20Changes%20in%20Urbanization.pdf World Bank Group.

Chapter 4

Study of Equatorial Plasma Bubbles Using ASI and GPS Systems

Dada P. Nade, Swapnil S. Potdar and Rani P. Pawar

Abstract

The plasma irregularities have been frequently observed in the F-region, at low latitude regions, due to the instability processes occurring in the ionosphere. The depletions in electron density, as compared to the background density, is a signature of the plasma irregularities. These irregularities are also known as the "equatorial plasma bubble" (EPB). These EPBs can measure by the total electron content (TEC) using GPS receiver and by images of the nightglow OI 630.0 nm emissions using all sky imager (ASI). The current chapter is based on the review on the signature of the EPBs in TEC and ASI. measurements. We have also discussed the importance of the study of EPBs.

Keywords: EPB, plasma irregularities, GPS and TEC

1. Introduction

The ionospheric radio wave communication, especially navigation is strongly influenced by the equatorial spread-F irregularities. Therefore, it is a scientific interest to do study these irregularities and its morphology and dynamic for the better communication system. The equatorial spread-F irregularities have been studied by several investigators using number of measurement techniques (e.g., [1–3]). The ground-based measurement techniques such as all sky imager (ASI), scanning photometer, RADAR and ionosonde are mostly used to study the dynamics of these irregularities. In equatorial-low latitude F-region, the electron density is depletes or enhance with respect to the background density due to vertical dynamics of F region. This is a key factor for generation of ionospheric irregularities. The size of these domains of irregularities ranges from a few hundred of kilometers in the east-west direction [4, 5] and a thousands of kilometers in north-south direction aligned with magnetic field lines (e.g., [6]). Basically, they occurred in the low latitude F region over an altitude 250–350 km, such irregularities have observed as dark and bright patches in the nightglow images of OI 630.0 nm emission. The dark patches are also called as equatorial plasma bubbles (EPBs) (e.g., [7, 8]) and bright patches are called as plasma blobs (e.g., [9, 10]).

Few investigators have been reported the nocturnal variations in occurrence (e.g., [7, 11, 12]) and zonal drift velocity (e.g., [8, 13]) of EPBs using OI 630.0 nm images. Many researchers have addressed generation mechanism of EPBs using different techniques (e.g., [14–16]). According to them, the EPBs are generated in the bottom side of F-region (e.g., [9, 17]) at the equator, mainly by the nonlinear evolution of the generalized Rayleigh-Taylor (GRT) instability [18] and E × B drift [14]. After sunset, the rapid uplifting of F-region is one of the key factors in generating of EPBs [19, 20].

The first time observation of plasma blobs has made with images of OI 630.0 nm, which were taken by the ground-based ASI from Cachoeira Paulista [21]. Using the similar data from the Boston University, Arecibo, Krall et al. [22] have also reported the observations of plasma blobs. The general characteristics of plasma blobs have been studied by using measurements of total electron content (TEC) [23].

Nade et al. [10], suggested that the plasma blobs occurred with EPBs and the variations in apex height may responsible for occurrence of the plasma blobs over low latitude region. However, the generation mechanism of EPBs and plasma blobs is not yet clearly understood [24] and it is most challenging issue to study the dynamics of the low latitude F region. In addition, several questions were raised by Kil et al. [25] on the EPB-blob connection. They reported that the creation of plasma blobs is not depending on EPBs and they also mentioned that the occurrence rate of EPBs varies with solar activity while blobs occurred frequently with solar activity. Based on these results they raised the question, why the occurrence of plasma blobs shows opposite nature with solar activity? These questions are creating inspiration to do more research in the same area.

2. Methodology

2.1 Nightglow emission OI 630.0 nm measurement

Based on numerous ground-based and in-situ studies, it is widely accepted that the nightglow OI 630.0 nm emissions are generated at low latitude F-region heights (250–300 km). The nightglow emission in F-region at (1D) 630.0 nm is governed by dissociative recombination between ions and electrons [26, 27]. The nightglow OI 630.0 nm images are used to study the characteristics of EPBs. Otsuka et al. [28] suggested that the ASI is an important aide towards improving the understanding of the coupling between ionosphere and thermosphere using images of nightglow OI 630.0 nm emission and OH emission. Because OH emissions are generating at around 100 km (ionosphere) while OI 630.0 nm emissions are generating at around 250 km (thermosphere).

Few methods are available in literature to analyze the all sky image data [29, 30]. Kubota et al. [31] has introduced a new method to convert the pixel images of the ASI into actual geographic coordinates for 250 km altitude, the airglow emission layer. Then to retrieve information from image data, the pixel value of images converted into corresponding latitude-longitude values by Narayanan et al. [32]. Recently, by the combination of both methods, Sharma et al. [33] has introduced the "average method" to process and analyze the image data.

Figure 1 illustrates processed images of OI 630.0 nm, which are taken on 17-18 January, 2012, showing the time evolution and structure of the EPBs and plasma blobs. In **Figure 1**, yellow and white arrows are showing signature of EPB and plasma blobs respectively in the OI 630.0 nm images.

To retrieve the pixel intensity from images of nightglow OI 630.0 nm Taori et al. [34] has introduced "the image crop method." In this method, we have selected only a square bin of 5 × 5 pixel at the center of the image corresponding to a rectangular field of view having ~1° along the zenith. The average intensity of this square bin is considered as intensity of OI 630.0 nm emission to study the nocturnal variation in

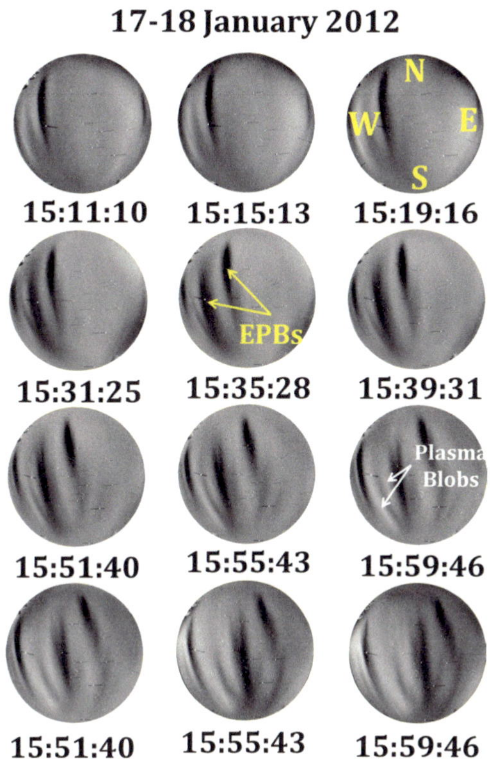

Figure 1.
Sequences of images of OI 630.0 nm obtained on 17-18 January, 2012 in IST at Kolhapur. Yellow and white arrows are showing dark and bright (EPB and plasma blob) respective structures in images.

intensity. Thus, we got the intensity data of OI 630.0 nm emission for each individual image with respect to time.

2.2 TEC measurements

Up to 11 GPS satellites are in view and provide outputs in 22 receiver channels [35]. The ionosphere has an effect on the signal of GPS satellite. TEC is measured along the path from the GPS satellite to a receiver. The TEC is defined by the integral of electron density in TEC unit (TECU), where $1\ TEC\ unit = 10^{16} electrons/m^2$ column along the signal transmission path. The dual frequency GPS receivers are used to measure the TEC, which is one of the most important methods to investigate the dynamics of ionosphere. Several research groups are showing interest in the equatorial ionospheric research using GPS data. Dow et al. [36] did an analysis of the importance of GPS data in support of the terrestrial reference frame, earth observations and research, positioning, navigation and timing as well as other applications that benefit the society. The slant TEC is the measure of the total number of free electrons in a column of unit cross section along the path of the electromagnetic wave between the satellite and the receiver [37].

A dual frequency (L1 = 1575.42 MHz and L2 = 1227.60 MHz) GPS receiver (LEICA GRX1200GGPRO GNSS) is operating at Hyderabad (17.37°N, 78.48°E) [3]. It is a unique station to study the ionospheric irregularities because it is located at the northern crest of the equatorial ionization anomaly (EIA). A dual-frequency GPS receiver can measure the difference in ionospheric delays between the L1 and L2 of the GPS frequencies, which are generally assumed to travel along the same path through the ionosphere. Thus, the group delay can be obtained as

$$\Delta(\delta t) = \delta t_{L1} - \delta t_{L2} \tag{1}$$

Here, $\Delta(\delta t)$ is a time delay in the pseudo-range (δt_{L1}) at L1 and pseudo-range (δt_{L1}) at L2. The resulting equation is (Jain et al., 2011),

$$\Delta(\delta t) = 40.3 \times TEC \times \frac{\left(f_{L1}^2 - f_{L2}^2\right)}{c \times f_{L1}^2 \times f_{L2}^2} \tag{2}$$

where f_{L1} and f_{L2} are the group path lengths corresponding to the high and low GPS frequencies $f_{L1} = 1575.42\ MHz$ and $f_{L2} = 1227.60\ MHz$, respectively and "c" is speed of light in vacuum. The TEC can be obtained by rewrite above equation as,

$$TEC = \frac{1}{40.3} \times \frac{c \times f_{L1}^2 \times f_{L2}^2}{\left(f_{L1}^2 - f_{L2}^2\right)} \times \Delta(\delta t) \tag{3}$$

The signal from different GPS satellites, at random elevation angles, recorded as a TEC measurements. These different satellites are identified by a pseudo-random number (PRN). The portions of the ionosphere cross by GPS signal depend on the elevation angle of GPS satellite. Therefore, in the present work the TEC data of only those GPS satellites, having elevation angles above 30° to avoid the multipath effect of signals, are considered. The maximum elevation angle over Hyderabad station is 60°. The STEC is measured at every 30 s by the GPS receiver.

3. Equatorial plasma bubble by GPS

The study of the spatial and temporal progress of EPBs formed in the ionosphere has been carried out using two different techniques radio waves as well as optical imaging over the globe. The optical imaging techniques have a limited coverage area, but have high resolution, while the radio wave techniques have a wide coverage area but can have low resolution for the ionospheric studies. The all sky imager is widely used instrument for the optical imaging of plasma bubble while GPS receivers used to study the ionospheric irregularities using radio waves. The **Figure 2** illustrates the occurrence of EPBs as D1 and D2 in the TEC measurements. The nocturnal variation in TEC with respect to local time (Indian Standard Time) observed on April 1, 2011 and April 2, 2011. The EPBs in TEC is indicated by D1 and D2. The occurrence period of EPBs is indicated by rectangular in the **Figure 2**.

Nishioka et al. [38] did a comprehensive study of the occurrence of plasma bubbles using ground-based GPS receiver from dip equator stations. They have considered Data from 2000 to 2006 from a network of 23 GPS receivers such as network of International GNSS Service (IGS), a GPS network by the Japan Agency for Marine-Earth Science and Technology (JAMSTEC), and Scripps Orbit Permanent Array Center (SOPAC), etc. They found a different characteristic rate of EPB occurrence in different regions also the dependency of the occurrence on the solar activity was different among the regions. They concluded that the sunset time lag effect plays an important role for the monthly variation and two asymmetries which could not be explained with the sunset time lag scenario (1) asymmetry between two solstices and (2) asymmetry between two equinoxes. They also found that the plasma bubble occurrence was high and constant for a stations having height on the dip equator (HODE) was <700 km and it is began to decrease for stations having HODE was higher than 700 km and was almost zero for the stations having HODE

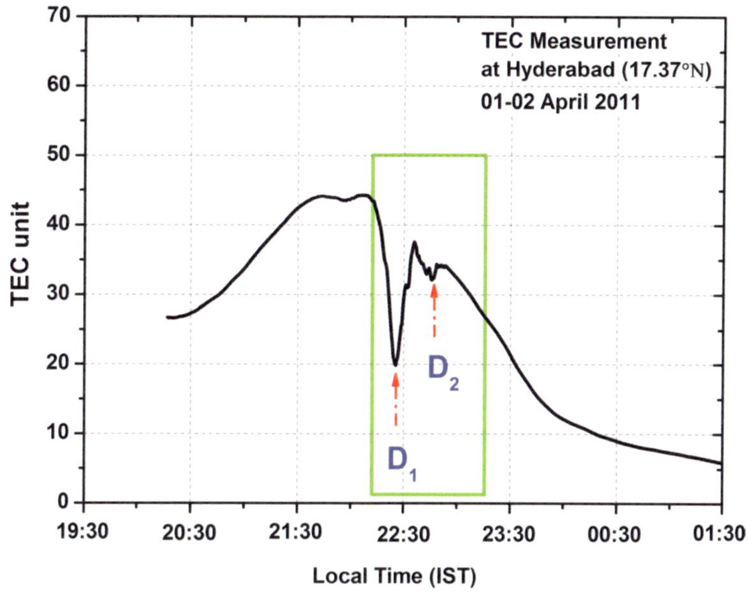

Figure 2.
The nocturnal variation in TEC with respect to local time (Indian Standard Time) observed on 01-02 April 2011. The EPBs in TEC is indicated by D_1 and D_2. The occurrence period of EPBs is indicated by rectangular.

higher than 900 km.. They defined HODE as shown in figure as an altitude of the geomagnetic field line on the magnetic dip equator which passes 400 km altitude above a site of GPS receiver.

Haase et al. [39] have studied the Propagation of plasma bubbles over Brazil from GPS and airglow data. They have mapped the airglow data to the GPS line-of-sight geometry for the direct comparison and revealing of resolvable westward tilt of the plasma depletion that may be due to vertical shear. They found the direct correspondence between integrated electron content (IEC) depletions and characteristics of depletions seen in horizontal airglow images, with very consistent observations of scale, amplitude, drift velocity and timing.

The EPB is monitored by using data provided by ground-based GNSS receiver Network over the South American continent by Takahashi et al. [40]. They have mapped the total electron content which could cover almost all of the continent within 4000 km distance in longitude and latitude. The TEC variability is monitored continuously with a time resolution of 10 min. The bubble structures are compared with simultaneous observations of OI630 nm all-sky image at Cachoeira Paulista (22.7°S, 45.0°W) and Cariri (7.4°S, 36.5°W). The formation and development of the bubble and eastward drifting features were successfully monitored and analyzed in this study. They found that the plasma bubbles observed during the December solstice has a periodic spacing, which is a periodic seeding mechanism of the bubbles.

The occurrence and characteristics of EPBs have been analyzed using the TEC data from GPS receivers over Hong Kong during 2001–2012 by Kumar et al. [41]. They found that the maximum occurrences of EPBs during the equinoctial months while minimum during the December solstice throughout 2001–2012. They also used the TEC data from different GNSS receivers over the Hong Kong. They concluded that the asymmetry in the EPB occurrences could be caused by the suppression of the growth rate of the instability by inter hemispheric neutral winds, which is known to be a primary cause for triggering EPB or ESF. The influence of solar and magnetic cycle is also studied.

Magdaleno et al. [42] studied the Climatology characterization of EPB using GPS data for the period 1998–2008. They have considered the slant total electron content (sTEC) derived from global positioning system (GPS) data from 67 International GNSS Service (IGS) stations distributed worldwide around the geomagnetic equator and the region of the ionospheric equatorial anomaly (IEA). The Ionospheric Bubble Seeker method is used to detect and distinguishes TEC depletions associated with EPBs. They found the largest occurrence rate of EPBs over the South America-Africa region and shown that the occurrence rate goes on decreasing as we go from the magnetic equator to higher latitudes.

First time study of the occurrence frequency of EPB over West Africa is done by Okoh et al. [42] using an ASI and GNSS Receivers from June 9, 2015 to January 31, 2017. This ASI is installed at Abuja (Geographic: 8.99°N, 7.38°E; Geomagnetic: 1.60°S) which covers almost the entire airspace of Nigeria. They found most occurrences of EPB during equinoxes and least occurrences during solstices also the occurrence rate of EPBs were highest around local midnight and lower for hours farther away. They also observed that the on/off status of EPB in airglow and GNSS observations are in 70% agreement.

Kumar [43] has studied the morphology of the EPB with respect of the solar activity over the Indian region from 2007 to 2012. The sTEC data are also considered from ground-based GPS receiver at Hyderabad (17.41° N, 78.55° E, Mag Lat 08.81° N) and two receivers at Bangalore (13.02°/13.03° N, 77.57°/77.51° E, Mag. Lat. 04.53°/04.55° N) in Indian region. He also observes that the occurrence of EPB is maximum in equinoctial months. He concluded that the equinox maximum in EPB occurrences for high solar activity years may be caused by the vertical F-layer drift due to pre-reversal electric field (PRE). This is expected to be maximum when daynight terminator aligns with the magnetic meridian, i.e., during the equinox months, whereas maximum occurrences during the solstice months of solar minimum could be caused by the seed perturbation in plasma density induced by gravity waves from tropospheric origins. The seasonal dependence of the EPBs occurrence is also studied.

Recently, Takahashi et al. [44] in detail studied the Occurrences of EPB (EPBs) and medium-scale traveling ionospheric disturbances (MSTIDs). They have used the GPS satellite data-based total electron content mapping, ionograms, and 630 nm all-sky airglow images observed over the South American continent during the period of 2014–2015. They observed a close relationship between the inter-bubble distance and the horizontal wavelength of the MSTIDs. They concluded that the MSTIDs are followed by EPBs primarily in the afternoon to the evening period due to the strong tropospheric convective activities (cold fronts and/or intertropical convergence zones) and the MSTIDs could be one of the seeding sources of EPBs.

Barros et al. [45] used ground-based network of GNSS receivers used to monitor EPB (EPBs) by mapping the total electron content (TEC map). They considered TEC data from GNSS receivers over South America for the period between November 2012 and January 2016. They found the latitudinal gradient varying from 123 ms^{-1} at the Equator to 65 ms^{-1} for 35° S latitude in the zonal drift velocities of the EPBs. They concluded that the latitudinal gradient in the inter-bubble distances seems to be related to the difference in the zonal drift velocity of the EPB from the Equator to middle latitudes and to the difference in the westward movement of the terminator.

Over the Thailand region, the statistical analysis of the separation distance between EPB is carried by Bumrungkit et al. [46]. The separation distance between EPBs is calculated using the Haversine formula technique in which the dual frequency GPS signal used. Their results show that the separation distances between EPBs on disturbed days in 2015 are in the range of 100–1200 km.

The effects of the R-T instability and EPB on the GPS signal are also studied by Panda et al. [47]. They have considered various instances of ionospheric disturbances triggered by natural processes such as earthquakes and volcanic eruption in the recent decade to investigate the spatiotemporal and seasonal effects of ionospheric irregularities on the GNSS signals. They found that the co-seismic ionospheric disturbances are difficult to study at the equatorial region due to mask of the EPBs but over the high latitude region these co-seismic ionospheric anomaly van be studied.

The effects of plasma bubbles on the GPS signal path and the positioning issue is studied by Moraes et al. [48]. The analysis focused on data from November 15, 2014 to November 30, 2014 and from February 4, 2015 to February 18, 2015, at São José DOS Campos, Brazil. They found that passing through the EPB, the radio signal may take a longer propagation path and have more losses of signal lock. The positioning errors may result in these cases.

Rajesh et al. [49] demonstrate that the EPBs appear to extend toward equator or pole as a result of the descending F layer and the recombination between free electrons of F layer and ions of the E layer at different latitudes. The apparent extension would vary from night to night depending on the post sunset vertical velocity of the F layer. Over equatorial region background electron density may be playing a vital role in providing the equinoctial asymmetry in the occurrence of ESF irregularities [16].

The effect of equatorial height variation of F region on ionospheric irregularities in the low latitude F region is also an important aspect. The apex height is also contributing in the occurrence of EPBs. Haaser et al. [50] suggested that EPBs occurred near the geomagnetic dip equator (<20° dip), typically within the nighttime ionospheric anomaly, while plasma blobs occurred mainly away from the geomagnetic dip equator, outside the anomaly regions (>15° dip). Yokoyama et al. [51] reported that the zonal structure of the plasma blobs in the northern hemisphere corresponded to that of the topside EPB in the southern hemisphere on a common magnetic flux tube, although the plasma blobs and the EPBs are separated by more than 20° in latitude.

Based on satellite data, Le et al. [52] reported that the localized eastward polarized electric field plays an important role in the creation of EPBs and plasma blobs. The strength of localized eastward polarized electric field is may depend on the virtual height of the F region at dip equator. They are directly proportional to each other. The fluctuations in the virtual height of the equatorial F region creates oscillations of F region in wave nature along the latitudes which is started from the magnetic equator to low latitude (±20°) crest of a wave. The background plasma density and fluctuations in plasma density are important factors for characterizing plasma blobs [53].

The eastward polarized electric field helps to combine flowing plasma with the background plasma over low latitude. Due to this combination, EPBs are generated over low latitude regions. According to literature serve, it is clear that ionospheric irregularities like EPBs or plasma blobs are not observed regularly. This may happen because of low combination or no combination of the plasma. Thus, this combination depends on the strength of eastward polarized electric field. If the strength of eastward polarized electric field is very low, then combinations will not possible while, for its particular value, low energy regions created which are captured in optical data as dark regions. These dark regions are also called as EPBs. If the strength of eastward polarized electric field is very high then, the rate of recombination will increase and high energy regions may create over low latitude regions. These high energy regions cause the enhancement in intensity of OI 630.0 nm emission, called as plasma blobs. The strength of eastward polarized electric field is also depends on the virtual height of F region at dip equator. Our results and analysis gives strong support for this statement. Herein, one more thing is that the flowing of plasma is in wavy nature, travel from the dip equator to low latitude. This nature of the flowing of plasma also makes an effect on the combination or recombination of plasma over low latitude regions. Thus to explain this nonlinear problem we need of more theoretical models.

4. Conclusions

Understanding of fundamental of EPBs, generation magnesium is very important for better communication and navigation system. Thus, we need more observations to study such phenomenon. In this chapter we have summarized the recent review on observation of EPBs using GPS and ASI.

Acknowledgements

The authors, DPN and SSP are highly indebted to Department of Science and Technology, Science and Engineering Research Board (DST-SERB), Government of India, for providing full financial assistance to carry out the present work under the major research project (File No. EEQ/2016/000275).

Author details

Dada P. Nade*, Swapnil S. Potdar and Rani P. Pawar
Department of Physics, Sanjay Ghodawat University, Kolhapur, India

*Address all correspondence to: dada.nade@gmail.com

References

[1] Abdu MA, Sobral JHA, Nelson OR, Batista IS. Solar cycle related range type spread-F occurrence characteristics over equatorial and low latitude stations in Brazil. Journal of Atmospheric and Terrestrial Physics. 1985;**47**(8-10):901-905

[2] Chapagain NP, Fejer BG, Chau JL. Climatology of postsunset equatorial spread F over Jicamarca. Journal of Geophysical Research: Space Physics. 2009;**114**(7)

[3] Nade DP, Shetti DJ, Sharma AK, Taori A, Chavan GA, Patil PT, et al. Geographical analysis of equatorial plasma bubbles by GPS and nightglow measurements. Advances in Space Research. 2015;**56**(9):1901-1910

[4] Sahai Y, Fagundes PR, Abalde JR, Pimenta AA, Bittencourt JA, Otsuka Y, et al. Generation of large-scale equatorial F-region plasma depletions during low range spread-F season. Annales de Geophysique. 2004;**22**(1):15-23

[5] Nade DP, Nikte SS, Ghodpage RN, Patil PT, Rokade MV, Vhatkar RS. Analysis of plasma bubbles observed in night airglow emission line OI 630.0 nm from Kolhapur using all sky imager. International Journal of Engineering Science Research. 2012;**03**(02):746-750

[6] Weber EJ, Buchau J, Moore JG. Airborne studies of equatorial F layer ionospheric irregularities. Journal of Geophysical Research: Space Physics. 2008;**85**(A9):4631-4641

[7] Sahai Y, Fagundes PR, Bittencourt JA, Abdu MA. Occurrence of large scale equatorial F-region plasma depletions during geo-magnetic disturbances. Journal of Atmospheric and Solar-Terrestrial Physics. 1998;**60**(16):1593-1604

[8] Nade DP, Sharma AK, Nikte SS, Patil PT, Ghodpage RN, Rokade MV, et al. Zonal velocity of the equatorial plasma bubbles over Kolhapur, India. Annales de Geophysique. 2013;**31**(11):2077-2084

[9] Pimenta AA, Fagundes PR, Bittencourt JA, Sahai Y, Gobbi D, Medeiros AF, et al. Ionospheric plasma bubble zonal drift: A methodology using OI 630 mn all-sky imaging systems. Advances in Space Research. 2001;**27**(6-7):1219-1224

[10] Nade DP, Sharma AK, Nikte SS, Chavan GA, Ghodpage RN, Patil PT, et al. Observations of plasma blobs by OI 630 nm using ASI and photometer over Kolhapur, India. Earth, Moon, and Planets. 2014;**112**(1-4):89-101

[11] Sahai Y, Fagundes PR, Bittencourt JA. Transequatorial F-region ionospheric plasma bubbles: Solar cycle effects. Journal of Atmospheric and Solar-Terrestrial Physics. 2002;**62**(15):1377-1383

[12] Sharma AK, Nade DP, Nikte SS, Patil PT, Ghodpage RN, Vhatkar RS, et al. Occurrence of equatorial plasma bubbles over Kolhapur. Advances in Space Research. 2014;**54**(3):435-442

[13] Fejer BG, Scherliess L, de Paula ER. Effects of the vertical plasma drift velocity on the generation and evolution of equatorial spread F. Journal of Geophysical Research: Space Physics. 2003;**104**(A9):19859-19869

[14] Kelley MC, Makela JJ, De La Beaujardiére O, Retterer J. Convective ionospheric storms: A review. Reviews of Geophysics. 2011;**49**

[15] Kakad B, Jeeva K, Nair KU, Bhattacharyya A. Magnetic activity linked generation of nighttime equatorial spread F irregularities. Journal of Geophysical Research: Space Physics. 2007;**112**(7)

[16] Sripathi S, Kakad B, Bhattacharyya A. Study of equinoctial asymmetry in the equatorial spread F (ESF) irregularities over Indian region using multi-instrument observations in the descending phase of solar cycle 23. Journal of Geophysical Research: Space Physics. 2011;**116**(11)

[17] Abalde JR, Fagundes PR, Sahai Y, Pillat VG, Pimenta AA, Bittencourt JA. Height-resolved ionospheric drifts at low latitudes from simultaneous OI 777.4 nm and OI 630.0 nm imaging observations. Journal of Geophysical Research: Space Physics. 2004;**109**(A11)

[18] Shvarts D, Alon U, Ofer D, McCrory RL, Verdon CP. Nonlinear evolution of multimode Rayleigh-Taylor instability in two and three dimensions. Physics of Plasmas. 1995;**2**(6):2465-2472

[19] Huang C, Lu Q, Wu M, Lu S, Wang S. Out-of-plane electron currents in magnetic islands formed during collisionless magnetic reconnection. Journal of Geophysical Research: Space Physics. 2013;**118**(3):991-996

[20] Sidorova LN, Filippov SV. Plasma bubbles in the topside ionosphere: Estimations of the survival possibility. Journal of Atmospheric and Solar-Terrestrial Physics. 2014;**119**:35-41

[21] Pimenta AA, Sahai Y, Bittencourt JA, Rich FJ. Ionospheric plasma blobs observed by OI 630 nm all-sky imaging in the Brazilian tropical sector during the major geomagnetic storm of April 6-7, 2000. Geophysical Research Letters. 2007;**34**(2)

[22] Krall J, Huba JD, Ossakow SL, Joyce G. Equatorial spread F fossil plumes. Annales de Geophysique. 2010;**28**(11):2059-2069

[23] Chen Y, Liu L, Le H. Solar activity variations of nighttime ionospheric peak electron density. Journal of Geophysical Research: Space Physics. 2008;**113**(11)

[24] Choi HS, Kil H, Kwak YS, Park YD, Cho KS. Comparison of the bubble and blob distributions during the solar minimum. Journal of Geophysical Research: Space Physics. 2012;**117**(4)

[25] Kil H, Choi HS, Heelis RA, Paxton LJ, Coley WR, Miller ES. Onset conditions of bubbles and blobs: A case study on 2 March 2009. Geophysical Research Letters. 2011;**38**(6)

[26] Bates DR. Emission of forbidden red and green lines of atomic oxygen from the nocturnal F region. Planetary and Space Science. 1992;**40**(7):893-899

[27] Vlasov MN, Nicolls MJ, Kelley MC, Smith SM, Aponte N, Gonzalez SA. Modeling of airglow and ionospheric parameters at Arecibo during quiet and disturbed periods in October 2002. Journal of Geophysical Research: Space Physics. 2005;**110**(A7)

[28] Otsuka Y, Shiokawa K, Ogawa T, Yokoyama T, Yamamoto M, Fukao S. Spatial relationship of equatorial plasma bubbles and field-aligned irregularities observed with an all-sky airglow imager and the equatorial atmosphere radar. Geophysical Research Letters. 2004;**31**(20)

[29] Garcia FJ, Taylor MJ, Kelley MC. Two-dimensional spectral analysis of mesospheric airglow image data. Applied Optics. 2008;**36**(29):7374

[30] Sekar R, Chakrabarty D, Sarkhel S, Patra AK, Devasia CV, Kelley MC. Identification of active fossil bubbles based on coordinated VHF radar and airglow measurements. Annales de Geophysique. 2007;**25**(10):2099-2102

[31] Kubota M, Fukunishi H, Okano S. Characteristics of medium- and large-scale TIDs over Japan derived from OI 630-nm nightglow observation. Earth, Planets and Space. 2001;**53**(7):741-751

[32] Lakshmi Narayanan V, Gurubaran S, Emperumal K. Imaging observations of upper mesospheric nightglow emissions from Tirunelveli (8.7°N). Indian Journal of Radio & Space Physics. 2009;**38**(3):150-158

[33] Sharma AK, Nade DP, Nikte SS, Ghodpage RN, Patil PT, Rokade MV, et al. Development of fast image analysis technique for all-sky images. Current Science. 2014;**106**(8):1085-1093

[34] Taori A, Jayaraman A, Kamalakar V. Imaging of mesosphere-thermosphere airglow emissions over Gadanki (13.5°N, 79.2°E) - first results. Journal of Atmospheric and Solar-Terrestrial Physics. 2013;**93**:21-28

[35] Dashora N, Pandey R. Observations in equatorial anomaly region of total electron content enhancements and depletions. Annales de Geophysique. 2005;**23**(7):2449-2456

[36] Dow JM, Neilan RE, Rizos C. The international GNSS service in a changing landscape of global navigation satellite systems. Journal of Geodesy. 2009;**83**(3-4):191-198

[37] Browne IC, Evans JV, Hargreaves JK, Murray WAS. Radio echoes from the moon. Proceedings of the Physical Society Section B. 1956;**69**(9):901-920

[38] Nishioka M, Saito A, Tsugawa T. Occurrence characteristics of plasma bubble derived from global ground-based GPS receiver networks. Journal of Geophysical Research: Space Physics. 2008;**113**(5)

[39] Haase JS, Dautermann T, Taylor MJ, Chapagain N, Calais E, Pautet D. Propagation of plasma bubbles observed in Brazil from GPS and airglow data. Advances in Space Research. 2011;**47**(10):1758-1776

[40] Takahashi H, Wrasse CM, Otsuka Y, Ivo A, Gomes V, Paulino I, et al. Plasma bubble monitoring by TEC map and 630nm airglow image. Journal of Atmospheric and Solar-Terrestrial Physics. 2015;**130-131**:151-158

[41] Kumar S, Chen W, Liu Z, Ji S. Effects of solar and geomagnetic activity on the occurrence of equatorial plasma bubbles over Hong Kong. Journal of Geophysical Research: Space Physics. 2016;**121**(9):9164-9178

[42] Okoh D, Rabiu B, Shiokawa K, Otsuka Y, Segun B, Falayi E, et al. First study on the occurrence frequency of equatorial plasma bubbles over West Africa using an all-sky airglow imager and GNSS receivers. Journal of Geophysical Research: Space Physics. 2017;**122**(12):12430-12444

[43] Kumar S. Morphology of equatorial plasma bubbles during low and high solar activity years over Indian sector. Astrophysics and Space Science. 2017;**362**(5)

[44] Takahashi H, Wrasse CM, Figueiredo CAOB, Barros D, Abdu MA, Otsuka Y, et al. Equatorial plasma bubble seeding by MSTIDs in the ionosphere. Progress in Earth and Planetary Science. 2018;**5**(1)

[45] Barros D, Takahashi H, Wrasse CM, Figueiredo CAOB. Characteristics of equatorial plasma bubbles observed by TEC map based on ground-based GNSS receivers over South America. Annales de Geophysique. 2018;**36**(1):91-100

[46] Bumrungkit A, Supnithi P, Saito S. Statistical analysis of separation distance between equatorial plasma bubbles near Suvarnabhumi International Airport, Thailand. Journal of Geophysical Research: Space Physics. 2018;**123**(9):7858-7870

[47] Panda D, Senapati B, Tyagi B, Kundu B. Effects of Rayleigh-Taylor instability and ionospheric plasma bubbles on the global navigation satellite

system signal. Journal of Asian Earth Sciences. 2019;**170**:225-233

[48] Moraes A d O, Vani BC, Costa E, Abdu MA, de Paula ER, Sousasantos J, et al. GPS availability and positioning issues when the signal paths are aligned with ionospheric plasma bubbles. GPS Solutions. 2018;**22**(4)

[49] Rajesh PK, Liu JY, Sinha HSS, Banerjee SB. Appearance and extension of airglow depletions. Journal of Geophysical Research: Space Physics. 2010;**115**(8)

[50] Haaser RA, Earle GD, Heelis RA, Klenzing J, Stoneback R, Coley WR, et al. Characteristics of low-latitude ionospheric depletions and enhancements during solar minimum. Journal of Geophysical Research: Space Physics. 2012;**117**(10)

[51] Yokoyama T, Su SY, Fukao S. Plasma blobs and irregularities concurrently observed by ROCSAT-1 and equatorial atmosphere radar. Journal of Geophysical Research: Space Physics. 2007;**112**(5)

[52] Le G, Huang CS, Pfaff RF, Su SY, Yeh HC, Heelis RA, et al. Plasma density enhancements associated with equatorial spread F: ROCSAT-1 and DMSP observations. Journal of Geophysical Research: Space Physics. 2003;**108**(A8)

[53] Kil H, Kwak YS, Lee WK, Krall J, Huba JD, Oh SJ. Nonmigrating tidal signature in the distributions of equatorial plasma bubbles and prereversal enhancement. Journal of Geophysical Research: Space Physics. 2015;**120**(4):3254-3262

Chapter 5

Spectral Optimization of Airborne Multispectral Camera for Land Cover Classification: Automatic Feature Selection and Spectral Band Clustering

Arnaud Le Bris, Nesrine Chehata, Xavier Briottet and Nicolas Paparoditis

Abstract

Hyperspectral imagery consists of hundreds of contiguous spectral bands. However, most of them are redundant. Thus a subset of well-chosen bands is generally sufficient for a specific problem, enabling to design adapted superspectral sensors dedicated to specific land cover classification. Related both to feature selection and extraction, spectral optimization identifies the most relevant band subset for specific applications, involving a band subset relevance score as well as a method to optimize it. This study first focuses on the choice of such relevance score. Several criteria are compared through both quantitative and qualitative analyses. To have a fair comparison, all tested criteria are compared to classic hyperspectral data sets using the same optimization heuristics: an incremental one to assess the impact of the number of selected bands and a stochastic one to obtain several possible good band subsets and to derive band importance measures out of intermediate good band subsets. Last, a specific approach is proposed to cope with the optimization of bandwidth. It consists in building a hierarchy of groups of adjacent bands, according to a score to decide which adjacent bands must be merged, before band selection is performed at the different levels of this hierarchy.

Keywords: hyperspectral, classification, band selection, spectral optimization, land cover

1. Introduction

High-dimensional remote sensing imagery, such as hyperspectral (HS) imagery, generates huge data volumes, consisting of hundreds of contiguous spectral bands. Several difficulties are caused by this high dimensionality. First, the Hughes phenomenon [1] can occur when classifying such data, even though modern classifiers such as support vector machines (SVM) and random forests (RF) are less sensitive to it [2, 3] except when very few training data are available [4]. Second, important computing times are required to process such high-dimensional data. Third, storing data requires huge volumes. Last, displaying high-dimensional imagery can be necessary, while human vision is limited to three colours [5, 6].

Hyperspectral data consist of hundreds of contiguous spectral bands, but most of these adjacent bands are highly correlated to each other. Thus a subset of well-chosen bands is generally sufficient for a specific problem. This enables to design adapted superspectral sensors dedicated to such specific land cover classification. **Spectral optimization** (SO) or optimal band extraction (BE) consists in identifying the most relevant spectral band subsets for such specific applications. Spectral optimization is a specific dimensionality reduction (DR). DR aims at reducing data volume minimizing the loss of useful information and especially of class separability. Dimensionality reduction techniques can be separated into **feature extraction** (FE) and **feature selection** (FS) categories.

FE consists in reformulating and summing up original information. Principal component analysis (PCA), minimum noise fraction (MNF), independent component analysis (ICA) and linear discriminant analysis (LDA) are examples of state-of-the-art feature extraction techniques. On the opposite, FS selects the most relevant features for a problem. When applied to HS data, it is named band selection (BS) and compared to FE; it enables to keep the physical meaning of the selected bands. For instance, in spectroscopy, FS has sometimes been performed by specialists identifying specific absorption bands or spectrum behaviour corresponding to a material, and this knowledge has then been used in expert systems (e.g. [7] for specific minerals, [8] for asbestos, [9] for asphalt or [10] for urban materials). At the end, SO is at the interplay between FS and FE as it aims at optimizing both band positions (FS) along the spectrum and width (FE).

This study aims at defining a SO strategy to design superspectral sensors dedicated to specific land cover classification problems. SO and FS are optimization problems involving both a **metric** (that is to say a score measuring the relevance of band subsets) to optimize and an **optimization** strategy. This study first focuses on the choice of a FS relevance score suitable for generic optimization heuristics. Both classification performance and selection stability will be considered. As an intermediate result, band importance profiles are considered providing hints about the relevance of the different parts of the spectrum. Once the FS criterion chosen, this chapter copes with the optimization of bandwidth, applying FS within a hierarchy of groups of adjacent bands.

2. FS: requirements and state of the art

In the state of the art, FS is often a first step in a specific classification workflow, while the context of this work is the design of superspectral sensors dedicated to specific land cover classification problems. Thus the selected band subset must be as efficient as possible **for most classifiers and not only for the used FS criteria**. Thus, their ability to discriminate between classes using selected feature subsets (that is to say their classification performance) independently from any classifier has to be considered to assess the FS criteria quality. Furthermore, the **stability** of the proposed solutions has also to be considered. Last but not least, in this sensor design context, constraints about the **maximum** number of bands to select exist. To sum it up, a good FS criterion for sensor design has to **be parsimonious, making it possible to select stable band subsets discriminant for most classifiers**. Thus, for a fair analysis, FS criteria must be compared for a same selected band subset size, and results must be evaluated according to different classifiers. Besides, computing time was not considered as an important criterion in this specific context of sensor design, where FS is not a preprocessing in a classification workflow.

Thus, this study focuses on the comparison of several FS **criteria** (presented in Section 2.1) for **supervised** classification problems (that is to say when classes and their ground truth are taken into account). To have a fair comparison, all these criteria will be optimized using the **same generic optimization** algorithms. It was here decided to use such generic optimization heuristics, in the context of sensor design, since such methods enable to easily control the number of bands to select and to add additional constraint within the band extraction process as in the second part of the study. The use of generic optimization methods necessarily excludes of the comparison feature ranking criteria (such as ReliefF [11, 12]) and FS methods where the score and the optimization method are strongly related to, for instance, SVM-RFE [13]. All criteria will be tested on several classic hyperspectral data sets.

2.1 FS: state of the art

Even though hybrid approaches involving several criteria exist [14, 15], FS methods and criteria are often differentiated between 'filter' (independent from any classifier), 'wrapper' (related to the classification performance of a classifier) and 'embedded' (related to the quality of classification models estimated by a classifier, but not directly to classification accuracy). It is also possible to distinguish supervised and unsupervised ones, especially for filters, that is to say whether a notion of classes is taken into account or not. All approaches mentioned below are summed up in **Tables 1** and **2**. Nevertheless, it must be kept in mind that hybrid approaches involving several criteria belonging to these different FS criteria categories often exist, as, for instance, in [14] or [15], where features are selected based on a wrapper method, respectively, guided or associated with filter criteria (mutual information between selected bands and between the ground truth).

2.1.1 Filter

Filter methods compute relevance scores independently from any classifier. Some filter methods are ranking approaches: features are ranked according to an individual score of importance. Such individual feature scores can be supervised or

Filters	
Unsupervised :	
Individual importance scores :	
PCA or ICA or between band correlation	[14, 16]
	[15, 17]
Feature subset scores :	
Mutual information	[18, 19]
Linear combination of bands	[20]
Other :	
SVC (Support Vector Clustering)	[9]
Method based on between band correlation	[15]
Supervised :	
Individual importance scores :	
ReliefF	[21, 22]
Other scores based on correlation or mutual information	[23]
Feature subset scores :	
Separability measures :	
Separability measures : Bhattacharyya	[8, 24]
or Jeffries-Matusita distances	[25, 26]
	[27]
Minimum Estimated Abundance Covariance	[28]
Correlation to ground truth :	
Mutual information	[2, 29, 30]
	[11, 31]

Wrapper	
Classification quality rates reached by :	
SVM classifier	[29, 32]
	[33, 34]
Maximum likelihood classifier	[27, 35]
Random Forests classifier	[36]
Target detector	[37]
Embedded	
Regularization models :	
L1-SVM	[38]
Lasso	[39, 40]
Other regularization models	[41, 42]
Built-in feature selection :	
Random Forests classifier	[6]
Decision Tree classifier	[7]
Individual feature importance scores :	
Importance measures from Random Forests	[6]
Multiple kernel SVM	[43]
Pruning methods :	
SVM-RFE	[44]
Generalization performance :	
SVM margin	[45, 46]
Out-Of-Bag error of a Random Forests classifier	[45, 46]
Other :	
Unmixing process	[47]

Table 1.
State of the art of feature selection criteria: the criteria that work with the FS criteria evaluation framework used in this study are underlined [16–47].

Filters
Unsupervised :
Individual importance scores : Pro : Fast. Against : Ranking selection, without considering redundancies between selected features. Solutions not parsimonious. Can be used to guide optimization. **Feature subset scores** and **linear band combinations :** Pro : Select the most representative and the less redundant features. Against : How to balance between these two criteria ? Against : Not necessarily the best solution (parsimonious/performance) for a classification problem.
Supervised :
Individual importance scores : Pro : Fast. Against : Ranking selection, without considering redundancies between selected features. Solutions not parsimonious. Can be used to guide optimization. **Feature subset scores :** **Separability measures :** Pro : Fast. Pro : Generally good solutions. Pro : Consider redundancies between selected features. Against : Classes modelled by an underlying parametric model (e.g. : Gaussian model). **Correlation to ground truth :** Pro : Fast. Pro : Select features that are the most correlated to the ground truth and the least redundant. Against : How to balance between these two criteria ?
Wrapper
Pro : Good results, especially for the classifier used to calculate the FS score. Pro : Consider (implicitly) both the redundancy between selected features and their classification performance. Against : Possible more important computing times depending on the used classifier Against : Risk to be dependent on the used classifier
Embedded
Regularization models : Pro : Good results. Pro : Fast. Pro : Consider redundancies between selected features. Against : More difficult to adapt them (e.g. to introduce additional other constraints) Against : Specific (not generic) optimization algorithm **Built-in feature selection :** Pro : Fast. Against : Not parsimonious. **Individual feature importance scores and pruning methods :** Pro : Fast. Against : Solutions not necessarily parsimonious (especially in case of a ranking FS). Against : Specific optimization method, not generic in case of pruning methods. **Generalization performance :** Pro : can be used like a wrapper while being faster. Pro : Consider (implicitly) redundancies between selected features. Against : Possible difference between the generalization performance and the classification performance of a model.

Table 2.
Pros and cons for the different families of FS criteria.

unsupervised. For instance, the well-known ReliefF score [11, 12] or scores measuring the correlation between features and ground truth [29] are supervised ones. However, such individual feature importance measures do not take into account the correlations between selected features. Thus, a feature subset composed of the n best features according to such measures is not necessarily an optimal solution, in the sense that it is not parsimonious.

Other ranking methods are unsupervised: they use importance measures calculated from a feature extraction technique. For instance, [48] ranks bands according to a score of importance calculated from PCA decomposition. Correlated bands are then removed according to a divergence measure. Du et al. and Hasanlou et al. [49, 50] have a similar approach using ICA instead of PCA. Other unsupervised approaches also use results of a PCA selecting the most similar features to the first PCA [46, 51].

Other filter approaches associate a score to feature subsets. In unsupervised case, [25] also performs a constrained energy minimization to select a set of bands having minimum correlation between each other. In supervised cases, separability measures such as Bhattacharyya or Jeffries-Matusita (JM) distances can be used in order to identify the best feature subsets for separating classes [30, 35, 45, 52]. Other separability measures based on the minimum estimated abundance covariance (related to the ability of the band subset to correctly unmix several sources) have also been used as in [53].

High-order statistics from information theory such as divergence, entropy and mutual information can also be used to select the best feature sets achieving the minimum redundancy and the maximum relevance, either in unsupervised situations as in [6, 22] or in supervised ones as in [14, 17, 54–56]. Martínez-Usó et al. [22] first clusters 'correlated' features and then selects the most representative feature of each group. Le Moan et al. [6] selects the three bands belonging to three red, green and blue spectral domains so that their correlation is minimized. In supervised cases, [14, 17, 54, 55, 57] select the set of bands that are more correlated to the ground truth and less correlated to each other. The most difficult is then to balance both criteria.

The orthogonal projection divergence [16] is another way to measure correlation between bands by the extent to which it is possible to express one band as a linear combination of the already selected bands. Last, [20] uses support vector clustering applied to features in order to identify the most relevant ones.

To sum it up, there are many various filter criteria corresponding to different approaches. Ranking methods according to an individual feature importance score remain limited, especially the ones only based on a supervised score, since they are not aware of the dependencies between selected features. Filter approaches associating a score to feature subsets are more interesting. Supervised and unsupervised approaches can be distinguished. Unsupervised approaches are interesting, but in a classification context, there is still a risk to select features that will not be all useful for the classification problem.

2.1.2 Wrapper

Wrapper relevance score associated with a feature set simply corresponds to its corresponding classification performance (measured by an accuracy score). Examples of such scores can be found in [14, 15, 58, 59] using SVM classifier, [60, 61] maximum likelihood classifier, [21] random forests, [46] spectral angle mapper or [26] a target detection algorithm.

2.1.3 Embedded

Embedded FS methods are also related to a classifier, but feature selection is performed using a feature relevance score different from a classification accuracy. Most of the time, embedded approaches directly select features during the classifier training step. Several types of embedded FS approaches can be distinguished [62].

Some embedded approaches are regularization based models. A classifier is trained according to an objective function where a fit-to-data term that minimizes the classification error is associated with a regularization function, penalizing models when the number of features increases or forcing model coefficients associated with some features to be small. Features with the coefficients close to 0 are eliminated. Examples of some approaches can be found in [23, 31, 63]. They also include the L1-SVM [64] and the least absolute shrinkage and selection operator (LASSO) FS [18, 63] approaches. Such approaches are fast and efficient. However, it can be more difficult to adapt them, for instance, to take into account additional constraints, since FS criterion and optimization method are linked.

Other embedded approaches use the built-in mechanism for feature selection in the training algorithm of some classifiers. For instance, random forests (RF) [41] and decision trees can be considered as performing an embedded feature selection, since, when splitting a tree node, only the most discriminative feature according to Gini impurity criterion is used among a feature subset randomly selected [41]. This FS eliminates the less useful features, but there is no guarantee to select a parsimonious feature subset: redundant features can be selected.

Some embedded approaches also provide feature importance measures, such as random forest classifier [41]. It is processed on samples left out of the bootstrapped samples and is based on the permutation decrease accuracy: the importance of a feature is estimated by randomly permuting all its values in these samples for each tree, as the difference averaged over all the trees between prediction accuracy before and after permuting this feature. Other embedded approaches providing feature importance use them in a pruning process that first uses all features to train a model, before progressively eliminating some of them while maintaining model performance. SVM-RFE [13] is a well-known embedded approach where the importance of the different features in a SVM model is considered. Such approach has been extended to multiple kernel SVM by [32], associating a different kernel to each feature, estimating the model and then using the weights associated with these kernels as feature importance measures.

Other approaches do not calculate a score of importance for each feature individually, but evaluate the relevance of sets of features. Such scores often measure the generalization performance of the obtained model. Thus, the FS is not directly performed during the training step, but uses an intermediate result of the training step. For instance, [37, 59] use the generalization performance, e.g. the margin of a SVM classifier, as a separability measure to rank sets of features. The out-of-bag error rate of a random forest [41] can also be considered as such score. These scores are calculated for feature subsets and measure the generalization performance of the model provided by the classifier. Thus, they can be considered as an alternative between filter separability measures and wrapper scores.

Embedded approaches can also be extended to unmixing methods, as, for instance, in [43] where band selection is integrated into an endmember and abundance determination algorithm by incorporating band weights and a band sparsity term into an objective function.

2.1.4 Optimization methods

Another issue for band selection is to determine the best set of features corresponding to a given criteria. An exhaustive search is often impossible, especially for wrapper techniques. Therefore, heuristics have been proposed to find a near-optimal solution without visiting the entire solution space. Optimization methods can be either specific to a FS method (as for most embedded ones) or generic. Generic optimization methods can be divided into two groups: sequential and stochastic.

Several incremental search strategies have been detailed in [44], including the sequential forward search (SFS) starting from one feature and incrementally adding another feature making it possible to obtain the best score or on the opposite the sequential backward search (SBS) starting for all possible features and incrementally removing the worst feature. Variants such as sequential forward floating search (SFFS) or sequential backward floating search (SBFS) are proposed in [44]. Serpico and Bruzzone [24] also proposes a variant of these methods called steepest ascent (SA) algorithms.

Among stochastic optimization strategies used for feature selection, several algorithms have been used for feature selection, including genetic algorithms [14, 15, 26, 37, 59], particle swarm optimization (PSO) [53, 58], clonal selection [61], ant colony [65] or even simulated annealing [30, 40].

In the specific case of hyperspectral data, adjacent bands are often very correlated to each other. Thus, hyperspectral band selection faces the problem of the clustering of the spectral bands. Band clustering/grouping has sometimes been performed in association with individual band selection. For instance, [15] first groups adjacent bands according to conditional mutual information and then performs band selection with the constraint that only one band can be selected per cluster. Su et al. [66] performs band clustering applying k-means to band correlation matrix and then iteratively removes the too inhomogeneous clusters and the bands that are too different from the representative of their cluster. Martínez-Usó et al. [22] first clusters 'correlated' features and then selects the most representative feature of each group, according to the mutual information. Chang et al. [40] performs band clustering using a more global criterion taking specifically into account the existence of several classes. Simulated annealing is used to maximise a cost function defined as the sum, over all clusters and over all classes, of correlation coefficients between bands belonging to a same cluster.

3. Which band selection criterion?

This study is a comparison of **FS criteria that can be optimized using generic optimization heuristics**, thus excluding several specific embedded or ranking approaches. The following FS criteria (listed in **Table 3**) were evaluated.

3.1 Compared FS criteria

3.1.1 Filter FS criteria

Filter criteria are independent from any classifier. Only scores assessing the relevance of feature subsets were considered, excluding filter FS methods ranking features independently according to an individual feature score (e.g. ReliefF).

3.1.1.1 Separability

Separability measures are used to identify the feature subsets achieving the best class distinction. Fisher, Bhattacharyya and Jeffries-Matusita measures [30, 35, 45, 52] are such scores. They were used assuming Gaussian class models. Let $\vec{\mu_i}$ and Σ_i be the mean and covariance matrices of the spectral distribution of class i. Fisher separability between classes i and j is defined in equation (1)

$$F_{ij} = \frac{\left(\vec{\mathbf{w}} \cdot \left(\vec{\mu_i} - \vec{\mu_j}\right)\right)^2}{{}^t\,\vec{\mathbf{w}}\,\left(\Sigma_i + \Sigma_j\right)\,\vec{\mathbf{w}}} \text{ where } \vec{\mathbf{w}} = \left(\Sigma_i + \Sigma_j\right)^{-1}\left(\vec{\mu}_i - \vec{\mu_j}\right) \tag{1}$$

Bhattacharyya separability between classes i and j is defined by equation 2.

Wrapper criteria	
svm.lin.K	Kappa of linear 1vs1 SVM
dt.K	Kappa of Decision Tree (DT)
rf.K	Kappa of Random Forests (RF)
ml.K	Kappa of Maximum Likelihood classifier (ML)
sam.K	Kappa of Spectral Angle Mapper (SAM)
sid.K	Kappa of Spectral Information Divergence (SID)
svm.lin.conf	confidence FS score of linear 1vs1 SVM
rf.conf	confidence FS score of Random Forests (RF)
Embedded criteria	
svm.lin.marg	Margin of a linear 1vs1 SVM
rf.oob	Out-of-Bag accuracy of Random Forest
Filter criteria	
fisher	Mean Fisher separability measure $= \frac{1}{nb\_pairs\_of\_classes} \sum_{i=1}^{c-1} \sum_{j=i+1}^{c} F_{ij}$
Bdist	Mean Bhattacharyya separability measure [8] $= \frac{1}{nb\_pairs\_of\_classes} \sum_{i=1}^{c-1} \sum_{j=i+1}^{c} B_{ij}$
jm	Jeffries-Matusita separability measure [8] $JM = \sum_{i=1}^{c-1} \sum_{j=i+1}^{c} (1 - e^{-B_{ij}})$
mi	Mutual Information based criterion [29] $J(S) = \sum_{f \in S} I(C,f) - \frac{1}{\#S} \sum_{f \in S} \sum_{s \in S; s \neq f} \frac{I(f,s)}{H(f).H(s)}$

Table 3.
Selected FS criteria to be compared.

$$B_{ij} = \frac{1}{8}^t \left(\vec{\mu_i} - \vec{\mu_j}\right) \Sigma^{-1} \left(\vec{\mu_i} - \vec{\mu_j}\right) + 0.5 \ln \left(\frac{\det\Sigma}{\sqrt{\det\Sigma_i \det\Sigma_j}}\right) \text{ where } \Sigma = \frac{\Sigma_i + \Sigma_j}{2} \tag{2}$$

As Bhattacharyya and Fisher separability measures are defined for binary problems, their mean overall possible pairs of classes were here used as FS criteria. To sum it up, the next separability measures were used as FS criteria:

- Mean Fisher (*fisher*) separability measures calculated over all pairs of classes (equation 3):

$$\frac{1}{\text{nb\_pairs\_of\_classes}} \sum_{i=1}^{c-1} \sum_{j=i+1}^{c} F_{ij} \tag{3}$$

- Mean Bhattacharyya (*Bdist*) separability measures calculated over all pairs of classes (equation 4):

$$\frac{1}{\text{nb\_pairs\_of\_classes}} \sum_{i=1}^{c-1} \sum_{j=i+1}^{c} B_{ij} \tag{4}$$

- Jeffries-Matusita measure (*jm*) defined in equation 5:

$$JM = \sum_{i=1}^{c-1} \sum_{j=i+1}^{c} \left(1 - e^{-B_{ij}}\right) \tag{5}$$

3.1.1.2 Mutual information

Another FS criterion based on high-order statistics from information theory, e.g. mutual information (MI), was adapted from [14] and tested: it took into account both feature-class dependencies and between feature correlations. It is defined in equation 6.

$$J(S) = \sum_{f \in S} I(C,f) - \frac{1}{\#S} \sum_{f \in S} \sum_{s \in S; s \neq f} \frac{I(f,s)}{H(f).H(s)} \tag{6}$$

for a feature subset S, where $I(C,f)$ is the MI between feature f and classes, $I(f,s)$ is the MI between features f and s and $H(f)$ is the entropy of feature f. It is referred to as *mi* in **Table 3**.

3.1.2 Wrapper and embedded FS criteria

Several classifiers were used to define wrapper scores related to their classification performances achieved using feature subsets. Only fast classifiers which did not require an optimization of hyper-parameters were used:

- **Maximum likelihood classification (ML)**: assuming a Gaussian model for the spectral distribution of classes, mean vectors and covariance matrices are estimated for each class during the training step. Each new sample is then labelled by its most probable class according to the model.

- **SAM and SID**: these classifiers are specific to hyperspectral data. The *spectral angle mapper* (SAM) consists in classifying a sample according to the angle between its spectrum and reference spectra. The *spectral information divergence* [42] comes from dissimilarity measures between statistical distributions and more precisely the Kullback-Leibler measure.

- **Support Vector Machine (SVM)** [67]: SVM has been intensively used to classify remote sensing data and especially hyperspectral data [2, 15, 28]. Training a SVM classifier aims at estimating the best frontiers between classes. Only a one-against-one linear SVM was used here. Indeed, it is fast and enables to avoid an optimization of hyper-parameters, contrary to other kernels. Besides, using a linear SVM introduces a constraint to select bands achieving a linear separation between classes.

- **Decision trees (DT)** [19].

- **Random forests (RF)** [41] is a modification of bagging applied with decision trees. It can achieve a classification accuracy comparable to boosting [41] or SVM [33]. It does not require assumptions on the distribution of the data, which is interesting when different types or scales of input features are used. It was successfully applied to remote sensing data such as multispectral data, hyperspectral data or multisource data. This ensemble classifier is a combination of tree predictors built from multiple bootstrapped training samples. For each node of a tree, a subset of features is randomly selected. Then, the best feature with regard to Gini impurity measure is used for node splitting. For classification, each tree gives a unit vote for the most popular class at each input instance, and the final label is determined by a majority vote of all trees.

These different classifiers were chosen because their underlying principles were different from each other. SAM, SID and ML rely on class models, while the others use inter-class separation models. RF can model even complex class frontiers remaining quite fast, while linear SVM selects features achieving the most possible linear separation between classes.

Wrapper FS scores measuring classification performance were considered:

- **Kappa coefficient**: for all of these classifiers, the *Kappa* coefficient has been used as a FS score.

- **Classification confidence score**: in addition, another FS score taking into account the classification confidence was also used [47]. Indeed, most classifiers provide classification confidence indices and a class membership measuring the degree to which the sample belongs to the different classes according to the classifier. Let $\mathcal{X} = \{(\mathbf{x}_i, y_i)\}_{1 \le i \le n}$ be a set of labelled ground truth samples x_i and their associated true label y_i. Let $m(x, c)$ be the class membership measuring the probability for $\mathbf{x}$ to belong to class c. A possible feature selection score $\mathcal{R}$ taking into account class membership measures and thus classification confidence can be defined by equation 7:

$$\mathcal{R}(\mathcal{X}) = \sum_{i=1}^{n} \delta(y_i, c(\mathbf{x}_i)).m(\mathbf{x}_i, c(\mathbf{x}_i)) \tag{7}$$

with $\delta(i,j) = \{-1 \text{ if } i \neq j \text{ and } 1 \text{ otherwise } \}$ and $c(\mathbf{x})$ the label given to $\mathbf{x}$ by the classifier. Such score measures both the ability to well classify the test samples for a given feature set and the separability between classes. Indeed, the more the samples are well classified, the more the score increases. The more the classifier is confident for well-classified samples, the more the score increases. The more the classifier is confident for bad-labelled samples, the more the score decreases. This confidence score was used in our experiments only for RF and linear SVM classifiers.

Embedded FS criteria. The two following criteria measuring the generalization performance of two classifiers were also tested. They are not pure embedded but can be considered as intermediate between wrapper and embedded. However, differentiating them from previous common wrapper scores, they are here referred to as 'embedded' in the sense that they assess the classification performance directly using a measure calculated directly while training the classifier and not after an evaluation of the model on a test data set. These scores are:

- The margin of a linear 1-vs-1 SVM classifier (without parameter optimization) (*svm.lin.marg*), that is to say the distance between the class frontier and its support vectors.

- The out-of-bag error [41] of a RF classifier (*rf.oob*). The out-of-bag samples are left out of the bootstrapped samples when training the RF.

3.2 Assessment approach

It must be kept in mind that study is a comparison of FS criteria and not of optimization methods. Thus all were optimized using the same optimization heuristics on the same classic hyperspectral data sets (3.3). The proposed workflow

(**Figure 1**) includes two steps. The suitable number of bands to select is first estimated for each data set, thanks to an incremental FS optimization algorithm called sequential forward floating search (SFFS) [44]. Then, the core comparison of FS criteria was performed. They were optimized to select this fixed number of bands using a stochastic FS optimization algorithm. A genetic algorithm (GA) (3.2.2) was used. Indeed, it proved to be efficient and generic enough to be used for all tested criteria. Besides it can provide valuable intermediate results (3.2.4) to assess FS stability. GA was launched several times to select this fixed number of bands for all tested FS criteria. It thus provided several possible band subset solutions. Indeed, performing FS several times was also a way to benefit from the stochastic nature of GA and thus to explore more band subset configurations. These different solutions were then quantitatively evaluated, according to different classifiers, to be able to draw conclusions about their relevance quite independently from a given classifier (3.2.3). Besides, to perform a qualitative analysis of the obtained solutions (and especially their stability), band importance measures were derived from intermediate results provided by this stochastic FS (3.2.2). It enabled to visually identify the parts of the spectrum considered as important by the FS criterion and to have a qualitative analysis concerning the stability of the proposed band subset solutions according to the FS criterion.

In practice, for each FS criterion, the GA feature selection process was launched five times on five limited data sets (100 training and 500 (300 for Indian Pines) testing samples) randomly selected with replacement among the whole data set. To sum it up, at the end, 25 'optimal' feature subset solutions were thus obtained for each criterion and had to be evaluated (**Figure 2**).

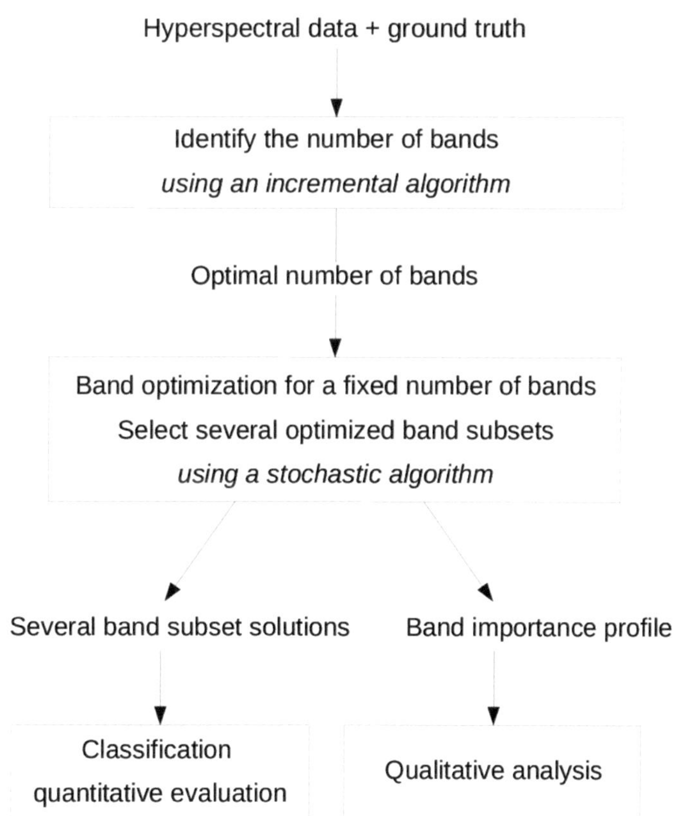

Figure 1.
Assessment process.

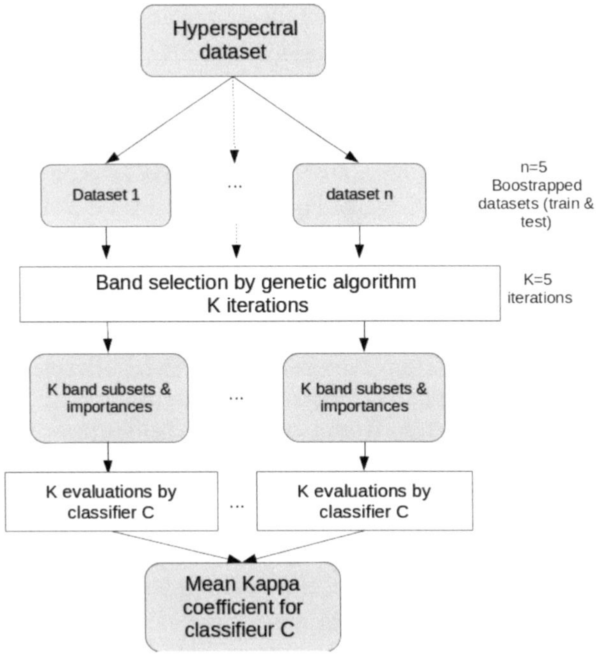

Figure 2.
Evaluation of FS criteria using band subsets obtained using a GA optimization.

3.2.1 Optimal band subset size using a sequential FS algorithm

Intermediate results of a sequential FS algorithm were used to identify how many bands must be selected. In our experiments, the sequential forward floating search (SFFS) algorithm was used [44].

This optimization method provides useful intermediate results. Indeed, it selects the 'best' sets of bands for different band subset sizes, starting from 1. Thus, it provides for each of them both the selected band subset (that could then be evaluated according to the performance of several classifiers) and the value reached by the FS score. Therefore, it enables to observe the evolution of FS score and classification quality, with the number of selected bands and then to decide how many bands are necessary to obtain suitable results. Other sequential methods as SVM-RFE [13] or SFS could also provide such information, but contrary to them, SFFS has the advantage to question at each step the selected set of bands obtained at the previous step, which enables possible modifications in the already selected band subset.

3.2.2 Band subset solutions using a genetic algorithm

Genetic algorithm (GA) is a family of stochastic optimization heuristics simulating the evolution mechanisms on a population of individuals. A score measuring its adaptation and its aptitude to stay alive is associated with each individual. In FS context, each individual is a feature subset and the score is the FS score.

Algorithm 1 Genetic algorithm.

It is intended to select less than p bands among a band set $\mathcal{B}$. J is the FS score to optimize.

Initialization: ($t \leftarrow 0$) Randomly generate a population $G(0)$ of N individuals, i.e. N sets of p bands.
while $t < t_{max}$ **do**
  //generation loop
  $t \leftarrow t+1$
  Calculate the score of each band subset of the current population.
  Keep only the n ($n < N$) best band subsets of the current population. Let $R(t)$ be this remaining population.
  Generate a new population $G(t)$ of N individuals from $R(t)$:
  for all new individual **do**
    Randomly select 2 parents among $R(t)$.
    Obtain a new individual by randomly crossing these 2 parents.
    Random mutations occur (randomly replacing a selected band by another one) in order to avoid to stay in a local optimum.
  end for
end while

3.2.2.1 GA-derived importance measures

The GA approach has some advantages for our problem. First, only the best solution is usually kept, while GA has visited many other candidates. Many of them have scores quite similar to the score of the best solution: they are almost as good as the final solution. Therefore, these intermediate results can be used to determine which bands are often selected in the solutions (see **Figure 3**) of these intermediate good band subset populations [27]. Thus, an individual band importance score $I(b)$ (defined in equation 8) is calculated for each band b, measuring the occurrence at which it has been selected by GA among the different n best sets of bands obtained for all generations

$$I(b) = \sum_{t} \sum_{R \in L(t)} \delta(b, R(t)) \text{ where } \delta(b, R) = 1 \text{ if } b \in R, \text{ 0 otherwise.} \tag{8}$$

To increase robustness, GA can be launched several times (i.e. so that different initializations and mutations occur) and over several training/testing sets randomly extracted from the whole data set. The proposed importance score is calculated for each of these results. Finally, the mean of these scores is considered for each band, giving the importance associated with this band.

3.2.3 Quantitative evaluation

In state of the art, FS is often considered as a first step in a specific classification workflow. In this context, wrappers are considered as achieving the best classification performance for a problem while sometimes lacking generality and being too classifier dependent. However, in our superspectral sensor design context, selected band subsets must be as efficient as possible for most classifiers and not only for the used FS criteria. Therefore, selected band subsets were here evaluated considering their classification quality reached with several classifiers.

Kappa coefficient was used as classification quality measure for the next classifiers: ML, RF and 1-vs-1 SVM with a radial basis function (RBF) kernel (with optimized parameters). It can here be noted that the latter was the only one not involved previously in a tested FS criterion. Thus, RBF SVM is the only classifier that is completely independent from all tested FS criteria. To come into details,

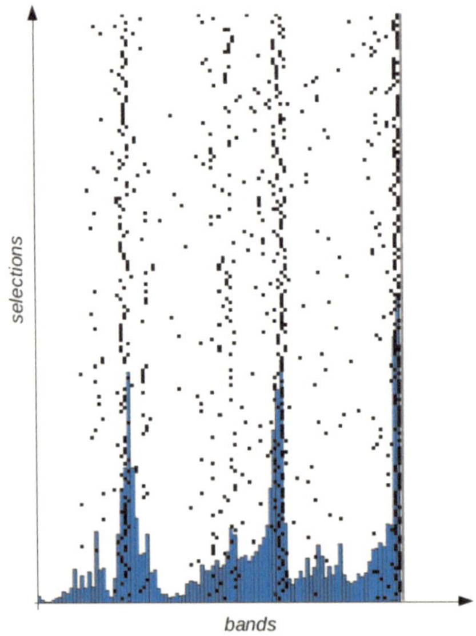

Figure 3.
Each line is a band subset selected in the intermediate results of GA, and each black dot represents a selected band. Blue histogram represents the importance associated with each band.

evaluation was performed and averaged on five training/testing sample sets: for each of them, classifiers were trained using 50 samples per class (in order to be in a difficult case with few training samples), and results were evaluated on all remaining ground truth samples. For each FS criterion, all selected band subsets (obtained for the several launches of the algorithm) were evaluated, and the mean Kappa coefficient was then computed over all of them (see **Figure 2**).

3.2.4 Selected band stability

Another evaluation criterion of the FS criteria quality was the stability of the selected features. As explained in Section 3.2.2, band importance profiles (**Figure 3**) can be derived from intermediate results of a GA feature selection. As the contiguous bands in hyperspectral data are correlated, such band importance profile should be quite regular and smooth (i.e. not too noisy). The smoothness/regularity of these profiles is thus related to the stability of the solutions obtained using a FS criterion. Furthermore, the final optimal solutions provided by the different launch of GA can also be examined. This analysis remains only qualitative.

3.3 Data sets

Three state-of-the-art available hyperspectral data sets were used for the experiments:

- **Pavia City Centre scene**[1]: This first data set is a hyperspectral scene acquired by the ROSIS sensor over the city centre of Pavia with a 1.3 m spatial resolution. It is a reflectance VNIR hyperspectral image with a

[1] Pavia data set is provided by Pavia University available at http://www.ehu.eus/ccwintco/index.php?title=Hyperspectral_Remote_Sensing_Scenes.

spectral resolution ranging from 460 nm to 860 nm. Noisy bands have been discarded, and only 102 spectral bands from the original 115 bands have been kept. It covers an urban area (city centre). Its associated land cover ground truth consists of nine urban classes (materials and vegetation).

- **Indian Pines scene**[2]: This hyperspectral scene was collected by the AVIRIS sensor over the Indian Pines test site in North-western Indiana. It is a radiance VNIR-SWIR hyperspectral image consisting of 220 spectral bands ranging from 400 to 2500 nm. Its associated ground truth consists of agricultural classes and other classes concerning perennial vegetation (forest, grass). In our experiments, only nine classes out of the original were kept. The discarded classes concerned less than 400 samples, which were considered as too few for our experiments.

- **Salinas scene**[3]: This hyperspectral scene was collected by the AVIRIS sensor over the Salinas Valley in California at a 3.7 m spatial resolution. It is an at-sensor radiance VNIR-SWIR hyperspectral image consisting of 224 spectral bands ranging from 400 to 2500 nm. Its associated ground truth consists of agricultural classes, that is to say different kinds of culture at different growing steps.

3.4 Results and discussion

3.4.1 Optimal number of bands using SFFS

An optimal number of bands to select was identified using SFFS incremental FS method, starting from one selected band and incrementing the band subset until a maximal number of bands. Indeed, this maximum number of bands was fixed to 20 considering the superspectral sensor design application, for which the number of possible spectral bands is limited. In practice, the influence of the number of selected bands on the FS score and on the classification performance (measured by Kappa and the F-score of the worst classified class) for a RBF SVM classifier using the best selected band subset was considered. The optimal number of bands was chosen as the one from which these scores virtually no longer increase. Results obtained using several FS scores were also considered to make this decision, and at the end, the number of bands to select is a trade-off between several FS criteria.

For Pavia data set, the influence of the number of selected bands on the FS score and on the classification performance (measured by Kappa and the F-score of the worst classified class) for a RBF SVM classifier using the best selected band subset can be seen in **Figure 4**. The different quality indices no longer evolve a lot from five bands, except the minimal F-score increasing slightly up to seven bands. Similar results were obtained using several FS criteria, even though some differences exist. For instance, the quality indices increased slower for *jm* than for *rf.conf* in **Figure 4**. Thus seven bands were selected for Pavia data set for further experiments.

---

[2] Indian Pines data set is provided by Purdue University and available at https://engineering.purdue.edu/~biehl/MultiSpec/hyperspectral.html.

[3] Salinas data set was downloaded from http://www.ehu.eus/ccwintco/index.php?title=Hyperspectral_Remote_Sensing_Scenes.

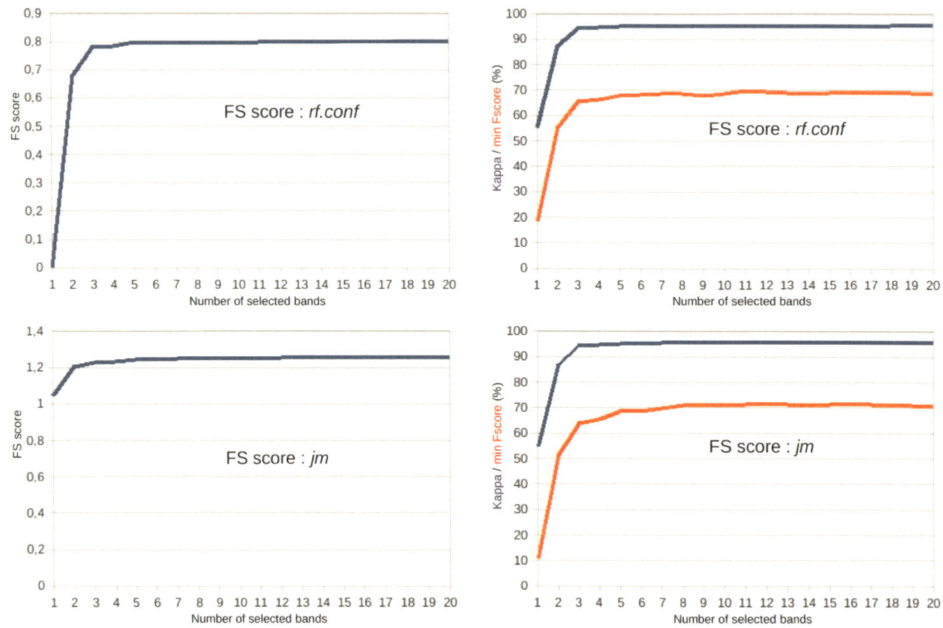

Figure 4.
Pavia test site: influence of the number of selected bands on the feature selection score (left) and on classification performance (using the best band subset with a RBF SVM classifier) (right with kappa coefficient for the blue line and F-score of the worst classified class for the red line). Two FS criteria tested: rf.conf (top) and jm (bottom).

The same kind of results was obtained for Salinas, and seven bands were also selected for this data set in further experiments.

For Indian Pines, obtained results are slightly different as shown in **Figure 5**. The FS score increases fastly until seven bands are selected. Then, it remains quite constant for *rf.conf* but continues to very slightly increase for *jm*. The same phenomenon can be observed for classification accuracies reached by a RBF SVM classifier using the selected band subsets. For *rf.conf* FS criterion, a maximum is reached around 10–11 selected bands, while for *jm*, a stage is reached for these values followed by a new slight increase.

However, it must be kept in mind that this data set is more difficult than the other ones. Indeed, on the one hand, it offers less training/testing samples (and thus an increased risk of over-fitting). On the other hand, classes are more difficult to distinguish to each other, and raw classification results (that is to say without any regularization post-processing step) remain noisy. Thus 10 bands were selected in further experiments for Indian Pines data set.

3.4.2 Comparison of FS criteria

GA optimization heuristic was then launched to select 7 bands for Pavia, 10 bands for Indian Pines and 7 bands for Salinas. For each FS score, several feature subset solutions were proposed using GA. Their classification quality rate (Kappa) (averaged over all of them) using several classifiers is presented in **Figure 6**. At the first glance, most of the time, Kappa coefficients reached using features selected according to different FS scores are correlated over the different classifiers (RBF SVM, RF and ML) used for evaluation. Indeed, if a FS score leads to the best classification for a classifier, it will also generally be the best for the other classifiers. Thus the relevance of score appeared to be quite independent from the classifier used at validation step.

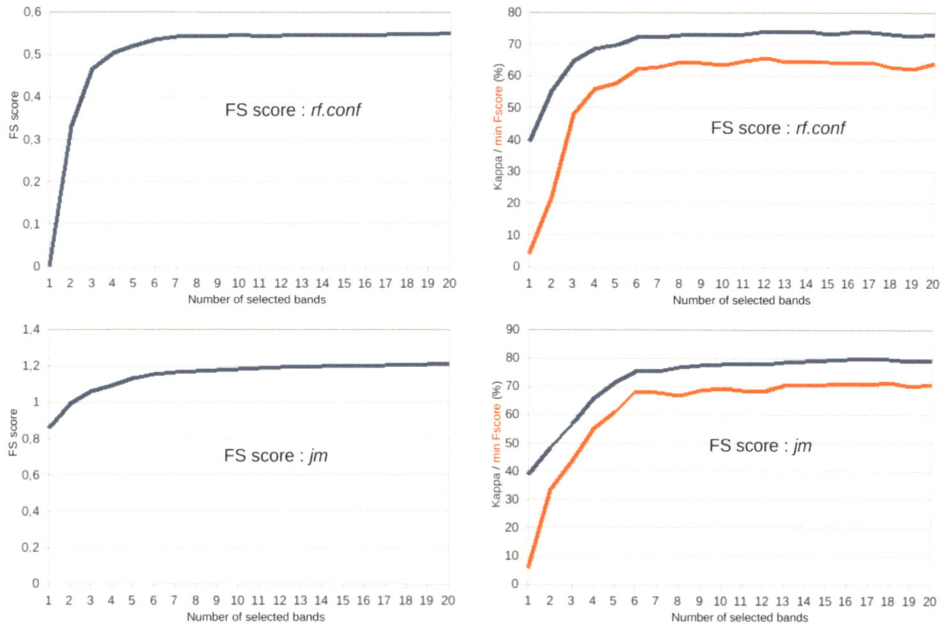

Figure 5.
Indian Pines test site: Influence of the number of selected bands on the feature selection score (left) and on classification performance (using the best band subset with a RBF SVM classifier) (right with kappa coefficient for the blue line and F-score of the worst classified class for the red line). Two FS criteria tested: rf.conf (top) and jm (bottom).

It can also be noticed from **Figure 6** that the best FS scores lead to quite equivalent classification quality. This is clearly visible for Pavia and to a less extent for Salinas. On the opposite, results are more contrasted on Indian Pines. This might be due to the fact that Indian Pines is a more difficult data set, with a stronger intra-class variability and inter-class similarity, whereas Pavia is a quite simple data set with few well-distinguished classes. These results will now be discussed for each category of FS criteria. Band importance provided by GA will also be considered.

3.4.2.1 Comparison of wrapper criteria

It can be seen from **Figure 6** that the FS scores *sam.K* and *sid.K* are less good than the other wrapper scores. This phenomenon appears strongly for Indian Pines and Salinas and is also a light trend for Pavia. The fact that it is more striking on Indian Pines scene can be related to the important intra-class variability of this data set.

The other wrapper scores relying on Kappa coefficient as a measure of classification performance lead to quite equivalent quantitative results. However, band importance profiles (**Figures 7** and **8**) provide other additional information. For instance, for Pavia data set (**Figure 7**), the FS score *svm.lin.K* tends to select the first bands (around band 5) of the spectrum, even though these bands are quite noisy. *ml.K* score performs very well considering classification performance but tends to be very sensitive to a probable atmospheric artefact, paying a lot of importance to bands from band 80 to band 85 and especially to band 82. This part of the spectrum corresponds to an atmospheric correction artefact, and not to a true discriminant phenomenon. This trend to select bands corresponding to this artefact is also observed for other FS scores.

Using classification confidence-based FS scores instead of classic classification accuracy scores tends to improve results. This trend can be observed in **Figure 6** both for RF and SVM: using *rf.conf* instead of *rf.K* or using *svm.lin.conf* instead of

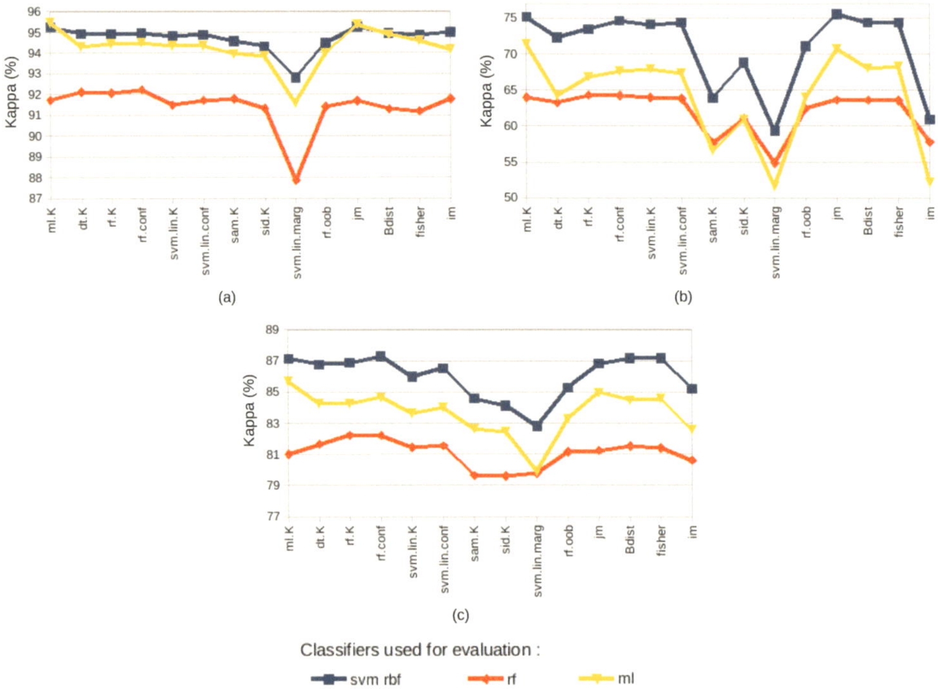

Figure 6.
Mean kappa coefficients obtained by classifiers RBF kernel SVM (red), RF (blue) and ML (yellow) using band subsets selected using the different FS criteria for the three data sets. From (a-c): Pavia, Indian Pines and Salinas.

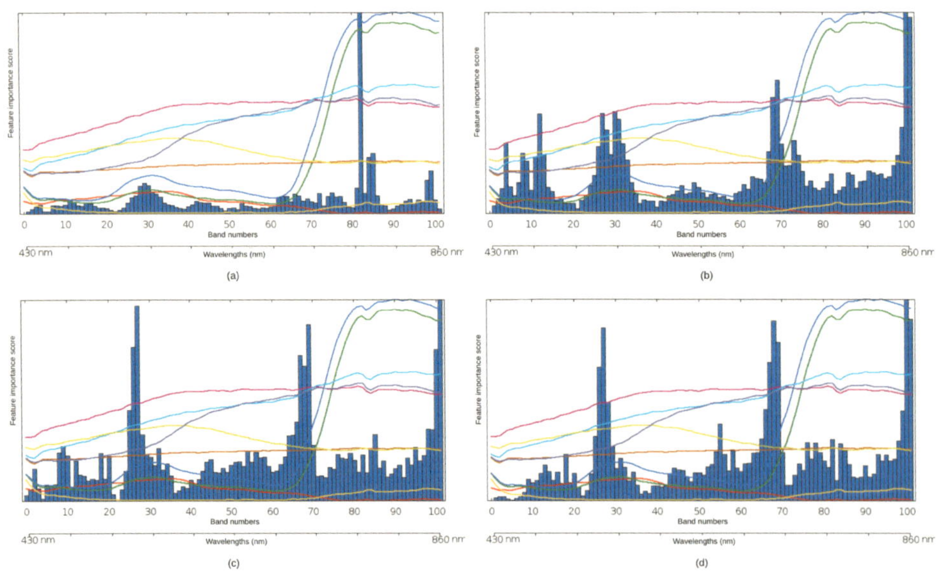

Figure 7.
Pavia test site: band importance profiles obtained using several FS criteria: (a) ml.K, (b) svm.lin.K, (c) rf.K and (d) rf.conf.

svm.conf tends to slightly improve classification quality. Considering band importance profiles obtained for Pavia (**Figure 7**), using *rf.conf* instead of *rf.K* avoids to select the noisy bands around band five. Band importance profiles obtained using *rf.conf* also seem to be slightly more regular than using *rf.K* both for Pavia (**Figure 7**) and Indian Pines (**Figure 8**). Thus, using a confidence-based FS score tends to regularize feature importances and thus to stabilize feature selection.

3.4.2.2 Comparison of wrapper and embedded criteria

Classification qualities reached using both tested embedded criteria (*svm.lin.marg* and *rf.oob*) appeared to be generally less good than using the wrapper scores associated with these two classifiers. This is especially clear for *svm.lin.marg*, which is the worst FS score, for all classifiers used at evaluation step.

Even though it performs quite well, feature subsets selected using *rf.oob* lead generally to worse classification performance than using the best wrapper scores, and especially *rf.K* and *rf.conf*, also associated to random forests.

3.4.2.3 Comparison of wrapper and filter criteria

Considering classification quality (**Figure 6**), mutual information (*mi*) leads to different results for the various data sets: on Pavia data set, feature subsets selected according to this FS score enable to reach classification performance as good as the best wrapper scores, while on Indian Pines data set, obtained results are among the worst. Band importance profiles (**Figures 9** and **10**) obtained using *mi* are also very different from those obtained for the other FS scores: they tend to neglect wide parts of the spectrum. This is especially striking for Indian Pines data set, where bands from 30 to 100 are not considered as important, contrary to other FS scores.

The other tested filter FS scores are separability measures. They perform very well considering classification quality (**Figure 6**): they lead to classification results as good or better than those obtained using the best wrapper FS scores. In particular, the Jeffries-Matusita separability distance (*jm*) appears to be one of the best FS scores.

However, considering band importance profiles obtained for Pavia (**Figure 9**) using *jm*, it tends to strongly focus on a part of the spectrum (bands 80 to 85) concerned by artefacts caused by atmospheric corrections. This phenomenon also

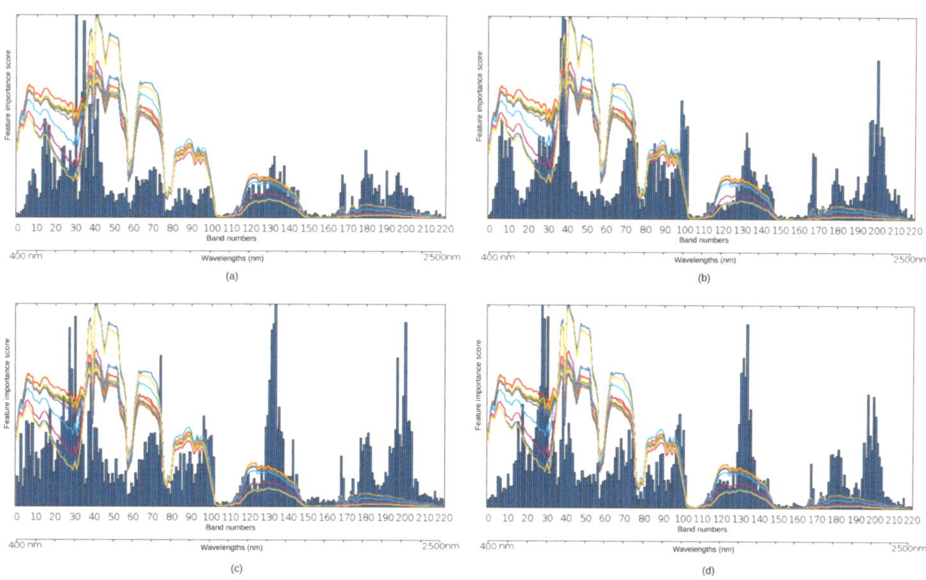

Figure 8.
Indian Pines test site: band importance profiles obtained using several FS criteria: (a) ml.K, (b) svm.lin.K, (c) rf.K and (d) rf.conf.

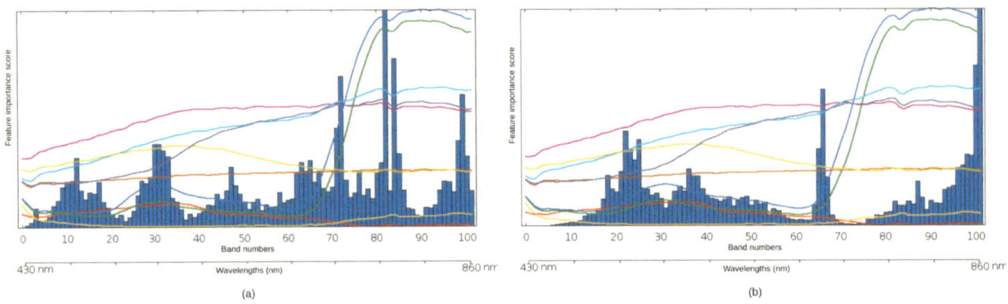

Figure 9.
Pavia test site. Band importance profiles obtained using several FS criteria: (a) JM distance and (b) mutual information.

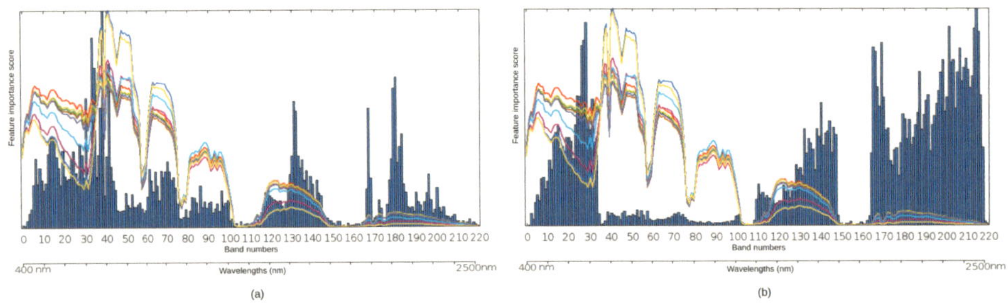

Figure 10.
Indian Pines test site. Band importance profiles obtained using several FS criteria: (a) JM distance and (b) mutual information.

occurred for *bdist* and *fisher* and, as explained above, was also observed for some wrapper FS scores.

Furthermore, band importance profiles obtained using *jm* FS score seem slightly more noisy or more difficult to interpret than using the best wrapper FS scores (*rf.K*,*rf.conf*).

3.4.3 Conclusion

FS score comparison. Some wrapper, embedded and filter FS scores were tested and evaluated on several data sets:

- *svm.lin.marg* appears clearly as the worst of them, performing poorly on all data sets.

- Other ones (*sam.K*, *sid.K* and *im*) perform quite good on simple data sets but poorly on the most difficult one (Indian Pines).

- Most perform well, leading to good classification performance. The best FS scores are filter separability measures or wrapper FS scores. However some slight trends can be observed:
    - Filter separability scores tend to lead to slightly better classification results than wrapper scores. Especially *jm* often appears as the best FS score according to quantitative analysis. However, considering band importance profiles, it tends to lead to less regular profiles and thus to less stable solutions than some wrapper scores. Besides they appear to be sensitive to an atmospheric correction artifact for Pavia data set.

- Confidence-based wrapper scores taking into account classification confidence (*rf.conf* or *svm.lin.conf*) perform better than classic wrapper scores expressed as a simple classification "hard label" error rate. This trend could be observed both in quantitative (classification performance) and qualitative (band importance profiles) analyses. Indeed, taking into account classification confidence tends to regularize feature importances and provide more stable feature subsets.

At the end, the most interesting FS scores are *rf.conf* for wrappers and *jm* for filters, since they lead to the best quantitative results. *rf.conf* seems to provide more stable results than *jm*, considering its more regularized band importance profile. Besides it is more robust to some artefacts (e.g. atmospheric correction artefact for Pavia). However, even though computing times were not discussed in this study, it must be added that FS selection using filter separability measures (such as *jm*) is faster than using wrapper scores such as *rf.conf*.

Thematic comments. Conclusions about interesting spectrum parts can be drawn using the importance profiles provided by the different FS criteria:

- Optimized spectral configurations are different from one FS criterion to another. Indeed, some parts of the spectrum are identified as important by most FS criteria, but other ones correspond to a clear disagreement.

- Spectrum parts considered as important can often be understood considering the spectra of classes. Indeed, they can correspond to almost constant spectrum parts located before or after a strong variation of spectra of some classes. They can also correspond to intersections between the spectra of several classes.

- For Indian Pines and Salinas scenes, no precaution was taken to handle noisy bands corresponding to the main atmospheric absorption windows. However, importance measures associated with these bands were very weak for most FS criteria (except the worse of them). Such observation can be considered as an additional quality criterion for the tested FS scores.

- Band importance profiles obtained for Indian Pines are often more difficult to analyse than for Pavia. Nevertheless, some common trends could be observed, especially in the SWIR domain, where some blobs along the spectrum are visible for most FS criteria and might correspond approximately to the locations of some spectral bands of the WorldView-3 satellite.

4. Exploring bandwidth and extracting optimal spectral bands using hierarchical band merging

Works in the previous section were dedicated to the identification of a FS score. It was used for band selection, that is to say to select a subset of original bands out of a hyperspectral data set (without optimizing their weights). This section will focus on band extraction and will consider band subsets composed of spectral bands with different spectral widths. Indeed, optimizing spectral width is important to design a spectral sensor, as having wider bands is a way to limit signal noise while having too wide bands can also lead to a loss a useful information.

4.1 Band grouping and band extraction: state of the art and proposed strategy

4.1.1 State of the art

Band grouping and clustering. In the specific case of hyperspectral data, adjacent bands are often very correlated to each other. Thus, band selection encounters the question of the clustering of the spectral bands of a hyperspectral data set. This can be a way to limit the band selection solution space. Band clustering/grouping has sometimes been performed in association with individual band selection. For instance, [15] first groups adjacent bands according to conditional mutual information and then performs band selection with the constraint that only one band can be selected per cluster. Su et al. [66] performs band clustering applying k-means to band correlation matrix and then iteratively removes the too inhomogeneous clusters and the bands too different from the representative of the cluster to which they belong. Martínez-Usó et al. [22] first clusters 'correlated' features and then selects the most representative feature of each group, according to mutual information. Chang et al. [40] performs band clustering using a more global criterion taking specifically into account the existence of several classes: simulated annealing is used to maximise a cost function defined as the sum, over all clusters and over all classes, of the sum of correlation coefficients between bands belonging to a same cluster. Bigdeli et al. and Prasad et al. [38, 68] perform band clustering, but not for band extraction: a multiple SVM classifier is defined, training one SVM classifier per cluster. Bigdeli et al. [68] has compared several band clustering/grouping methods, including k-means applied to the correlation matrix or an approach considering the local minima of mutual information between adjacent bands as cluster borders. Prasad and Bruce [38] proposes another band grouping strategy, starting from the first band of the spectrum and progressively growing it with adjacent bands until a stopping condition based on mutual information is reached.

Band extraction. Specific band grouping approaches have been proposed for spectral optimization. De Backer et al. [30] defines spectral bands by Gaussian windows along the spectrum and proposes a band extraction optimizing score based on a separability criterion (Bhattacharyya error bound) thanks to a simulated annealing. [34] merges bands according to a criteria based on mutual information. Jensen and Solberg [69] merges adjacent bands decomposing some reference spectra of several classes into piece-wise constant functions. Wiersma and Landgrebe [70] defines optimal band subsets using an analytical model considering spectra reconstruction errors. Serpico and Moser [52] proposes an adaptation of his steepest ascent algorithm to band extraction, also optimizing a JM separability measure. Minet et al. [26] applies genetic algorithms to define the most appropriate spectral bands for target detection. Last, some studies have also studied the impact of spectral resolution [71], without selecting an optimal band subset.

4.1.2 Proposed approach

The approach proposed in this study consists in first building a hierarchy of groups of adjacent bands. Then, band selection is performed at the different levels of this hierarchy.

Thus, it is here intended to use the hierarchy of groups of adjacent bands as a constraint for band extraction and a way to limit the number of possible combinations, contrary to some existing band extraction approaches such as [52] that extract optimal bands according to JM information using an adapted optimization method or [26] that directly use a genetic algorithm to optimize a wrapper score.

4.2 Hierarchical band merging

The first step of the proposed approach consists in building a hierarchy of groups of adjacent bands that are then merged. Even though it is here intended to be used to select an optimal band subset, this hierarchy of merged bands can also be a way to explore several band configurations with varying spectral resolution, that is to say with contiguous bands with different bandwidth.

4.2.1 Proposed algorithm

Notations. Let $B = \{\lambda_i\}_{0 \leq i \leq nbands}$ be the (ordered) set of original bands. Let $H = \{H^{(i)}\}_{0 \leq i < nlevels}$ be the hierarchy of merged bands. $H^{(i)} = \left\{H_j^{(i)}\right\}_{1 \leq j \leq n_i}$ is the ith level of this hierarchy of merged bands. It is composed of n_i merged bands, that is to say n_i ordered groups of adjacent bands from B.

Thus, each $H_j^{(i)}$ is defined as a spectral domain:

$$H_j^{(i)} = \left[H_j^{(i)}.\lambda_{min}; H_j^{(i)}.\lambda_{max}\right]$$

Thus, the merged band $B_1 \oplus B_2$ obtained when merging two such adjacent merged bands B_1 and B_2 is $B_1 \oplus B_2 = [B_1.\lambda_{min}; B_2.\lambda_{max}]$.

Let $J(.)$ be the score that has to be optimized during the band merging process.

The proposed hierarchical band merging approach is a bottom-up one. The algorithm is defined below:

Initialization: $H^{(0)} = B$ (that is to say that each merged band of the first level of the hierarchy only contains one individual original band).

Band merging: create level l + 1 from level l:

Find the pair of adjacent bands at level l that will optimize the score if they are merged: find $\hat{k} = argmin_k J(T(H^{(l)}, k))$ with $T(H^{(l)}, k) = \left[H_0^{(l)}; \ldots; H_{k-1}^{(l)}; H_k^{(l)} \oplus H_{k+1}^{(l)}; H_{k+2}^{(l)}; \ldots; H_{n_l}^{(l)}\right])$.

Then $H^{(l+1)} = T\left(H^{(l)}, \hat{k}\right)$.

A table L_l^{l+1} is defined to link the different merged bands at consecutive hierarchy levels:

$$\text{for } 1 \leq j \leq \hat{k}, L_l^{l+1}\left(H_j^{(l)}\right) = H_j^{(l+1)}$$

$$L_l^{l+1}\left(H\hat{k}^{(l)}\right) = H\hat{k}^{(l+1)}$$

$$L_l^{l+1}\left(H\hat{k}^{(l)} + 1\right) = H\hat{k}^{(l+1)}$$

$$\text{for } \hat{k} + 2 \leq j \leq n_l, L_l^{l+1}\left(H_j^{(l)}\right) = H_{j-1}^{(l+1)}.$$

At the end, the value of a pixel in a merged band is defined as the mean of its values over the different bands it contains.

4.2.2 Band merging criteria

Several optimization scores J were examined. (In the algorithm described in Section 4.2.1, this score is aimed to be minimized.) They can be either supervised or unsupervised, depending whether classes are considered or not at this step.

4.2.2.1 Correlation between bands

Between band correlation (either the classic normalized correlation coefficient or mutual information) (see **Figure 11**) measures the dependence between bands. So a first band merging criterion intends to merge adjacent bands considering how they are correlated to each other. Thus, it tries to obtain consistent groups of adjacent correlated bands.

Such measure inspired from [40] can be defined by the next function in equation 9 (intended to be minimized):

$$J\left(H^{(l)}\right) = \sum_{i=1}^{n_l} \sum_{b_1=H_i^{(l)}.\lambda_{min}}^{H_i^{(l)}.\lambda_{max}} \sum_{b_2=H_i^{(l)}.\lambda_{min}}^{H_i^{(l)}.\lambda_{max}} (1 - c(b1, b2)) \tag{9}$$

where $c(b_1, b_2)$ is the correlation score between bands b_1 and b_2.

4.2.2.2 Spectra approximation error

Band merging can also use the method as described in [69] to decompose some reference spectra of several classes into piece-wise constant functions (**Figure 12**). Adjacent bands are then merged trying to minimize the reconstruction error between the original and the piece-wise constant reconstructed spectra.

Such measure is defined by the next function (see equation 10) for a set $s_{j1 \le j \le n_s}$ of n_s spectra:

$$J\left(H^{(l)}\right) = \sum_{j=1}^{n_s} \sum_{i=1}^{n_l} \sum_{b=H_i^{(l)}.\lambda_{min}}^{H_i^{(l)}.\lambda_{max}} |s_j(b) - mean\left(s_j, H_i^{(l)}\right)| \tag{10}$$

where $mean\left(s_j, H_i^{(l)}\right)$ denotes the mean of spectra s_j over spectral domain $H_i^{(l)}$.

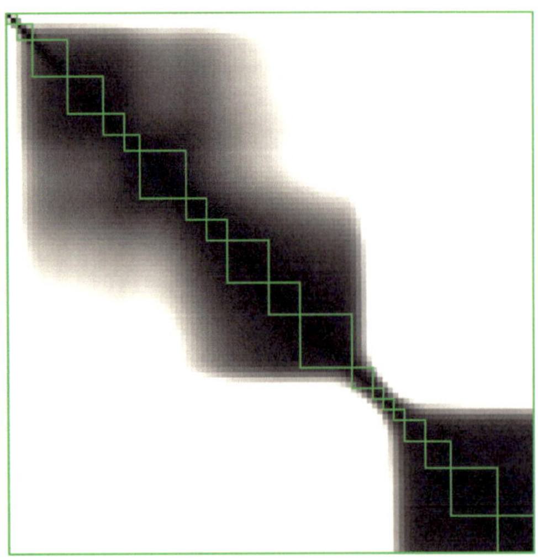

Figure 11.
Examples of groups of bands superimposed on the band correlation matrix (for Pavia data set).

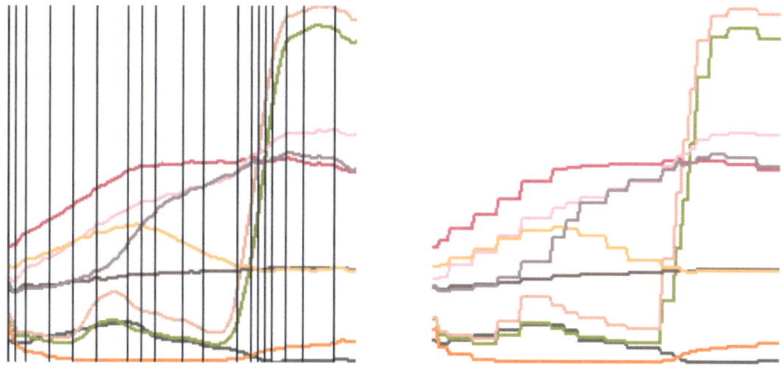

Figure 12.
On the left, examples of merged bands superimposed on the original reference spectra. On the right, piece-wise constant reconstructed spectra for these merged bands (Pavia data set).

4.2.2.3 Separability

Another criterion to merge adjacent band is their contribution to separability between classes. Possible separability measures are the Bhattacharyya distance (*B-distance*) or the Jeffries-Matusita distance [35, 52] already used as FS score in 3.

At a level of the band merging hierarchy, the best set of merged bands is the one that maximizes class separability. So a possible criterion J (to minimize) for band merging can be defined by equation 11 as

$$J\left(H^{(l)}\right) = -JM\left(H^{(l)}\right) \tag{11}$$

4.2.3 Results

Figure 13 shows results on Pavia data set for the three criteria described in the previous section. The separability-based criterion tends to lead to more different results than the other ones. The different criteria do not consider the same parts of the spectrum as having to be kept at fine resolution. For instance, correlation or spectra reconstruction criteria tend to fast merge bands between number 30 and 32, while separability tends to preserve them at fine resolution. On the opposite, separability tends to fast merge some bands in the red-edge domain, while the other criteria keep this domain at fine resolution. This can be understood considering the underlying criteria; indeed adjacent bands are not very correlated to each other in this domain, and the slope of spectra is strong for vegetation classes; thus they cannot be merged easily according to correlation or spectra approximation error band merging criteria. On the opposite, the only interesting information for classification (e.g. for class separability) is the fact there is a slope there and thus the values of the bands before and after this domain. Thus, merging these red-edge bands will have a little impact on class separability.

As the hierarchy of merged bands can also be a way to explore several band configurations with varying contiguous bands with different spectral resolution, the different band configurations corresponding to the different levels were evaluated using a classification quality measure. Thus, for each level, a classification was performed using a support vector machine (SVM) classifier with a radial basis function (rbf) kernel and evaluated. Its Kappa coefficient was considered.

Such results are presented on **Figure 14**. It can be seen that some spectral configurations made it possible to obtain better results than at original spectral resolution. Configurations obtained using the correlation coefficient are generally less good than for the two other criteria. Except for Pavia, the spectra piece-wise approximation

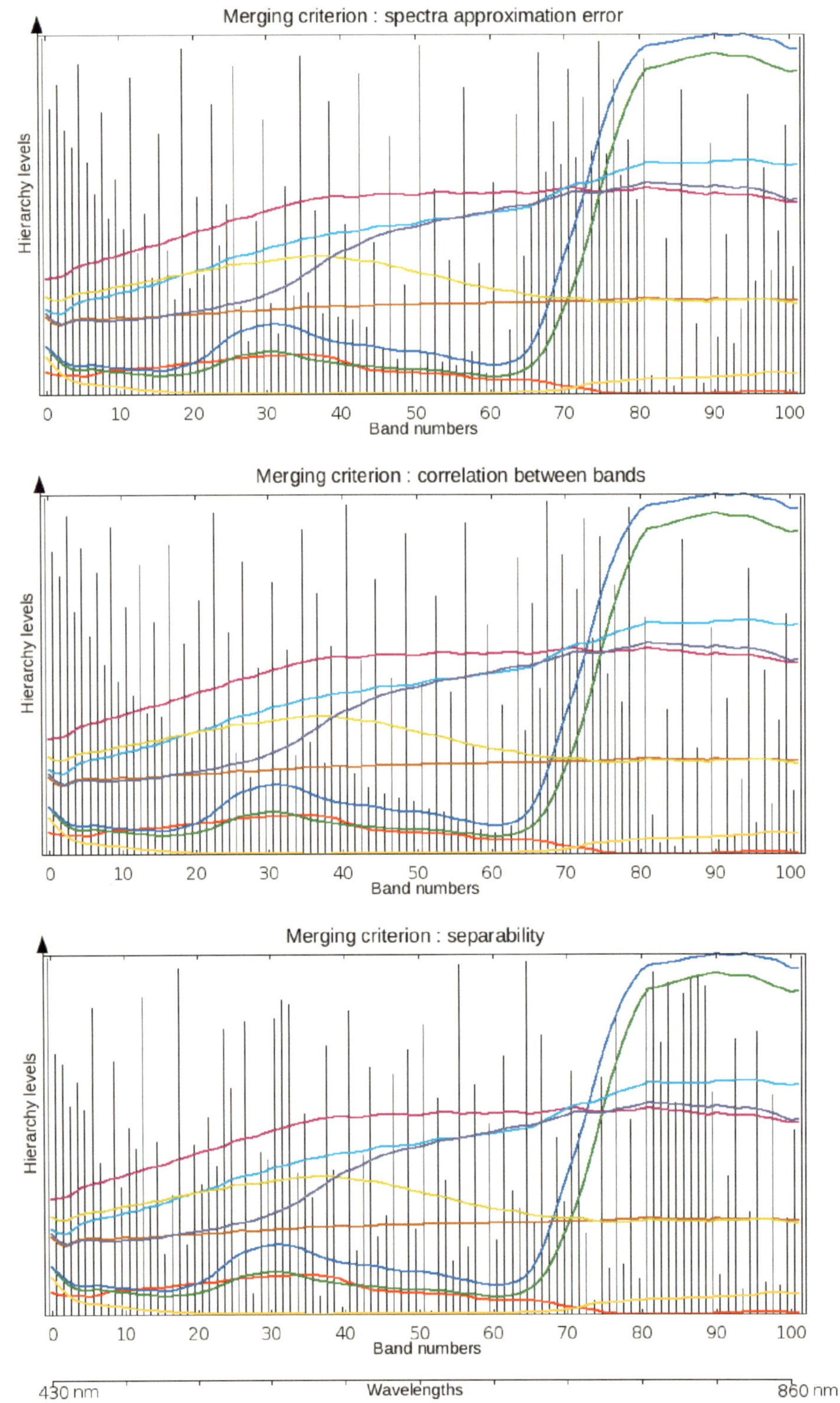

Figure 13.
Hierarchies of merged bands obtained for different criteria for Pavia data set: Spectra piece-wise approximation error (top), between band correlation (middle) and class separability (bottom). X-axis corresponds to the band numbers/wavelengths. y-axis corresponds to the level in the band merging hierarchy (bottom, finest level with original bands; top, only a single merged band). Vertical black lines are the limits between merged bands: the lower the hierarchy, the more the merged bands are. Reference spectra of the classes are displayed in colour.

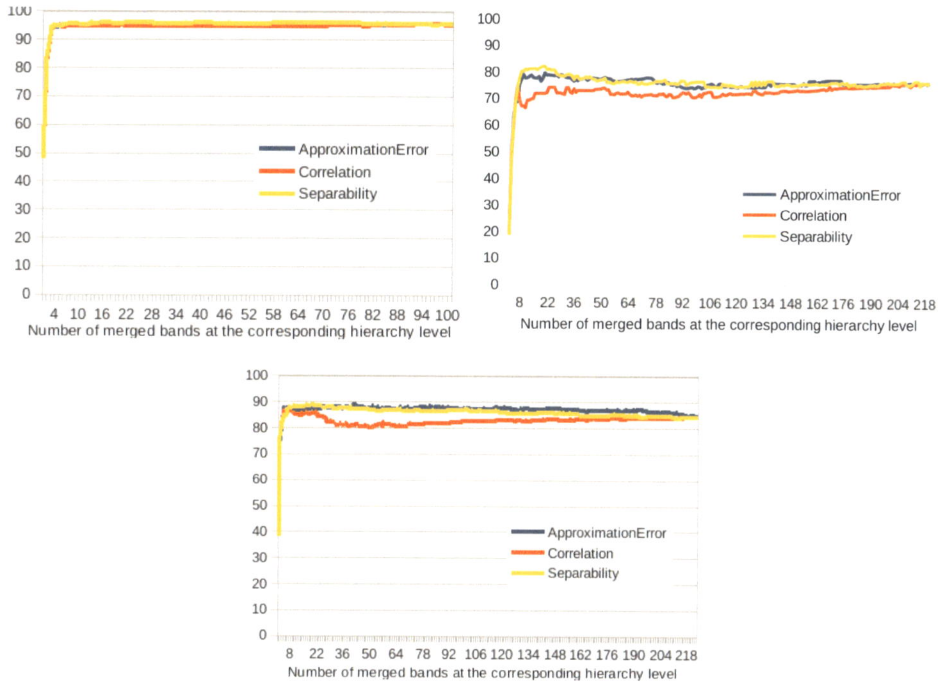

Figure 14.
Kappa (in %) reached by a rbf SVM for the different band configurations of the hierarchy (x-axis = number of merged bands in the spectral configuration corresponding to the hierarchy level): for Pavia (top), Indian Pines (middle) and Salinas (bottom) data sets.

error merging criterion tends to lead to the best results. But for Pavia, the classification Kappa reached using the different criteria remained very similar.

4.3 Band selection within the hierarchy

4.3.1 Greedy algorithm

To optimize spectral configuration for a limited number of merged bands, a greedy approach was first used: it performed band selection at the different levels of the hierarchy of merged bands, paying no attention at results obtained at the previous level. Thus a set of merged bands was selected at each level of the hierarchy.

The feature selection (FS) score to optimize was the JM separability measure. It was optimized at each level of the hierarchy using an SFFS incremental optimization heuristic [44].

4.3.1.1 Results

Obtained results on Pavia data set are presented on **Figure 15**: five merged bands (as in [27]) were selected at each level of the hierarchy of merged bands. The positions of the selected merged bands do not change a lot when climbing the hierarchy, except when reaching the lowest spectral resolution configurations. At some levels of the hierarchy, the position of some selected merged bands can also move and then come back to its initial position when climbing the hierarchy.

Thus, it can be possible to use the selected bands at a level l to initialize the algorithm at the next level $l + 1$. This modified method will be presented in Section 4.3.2.

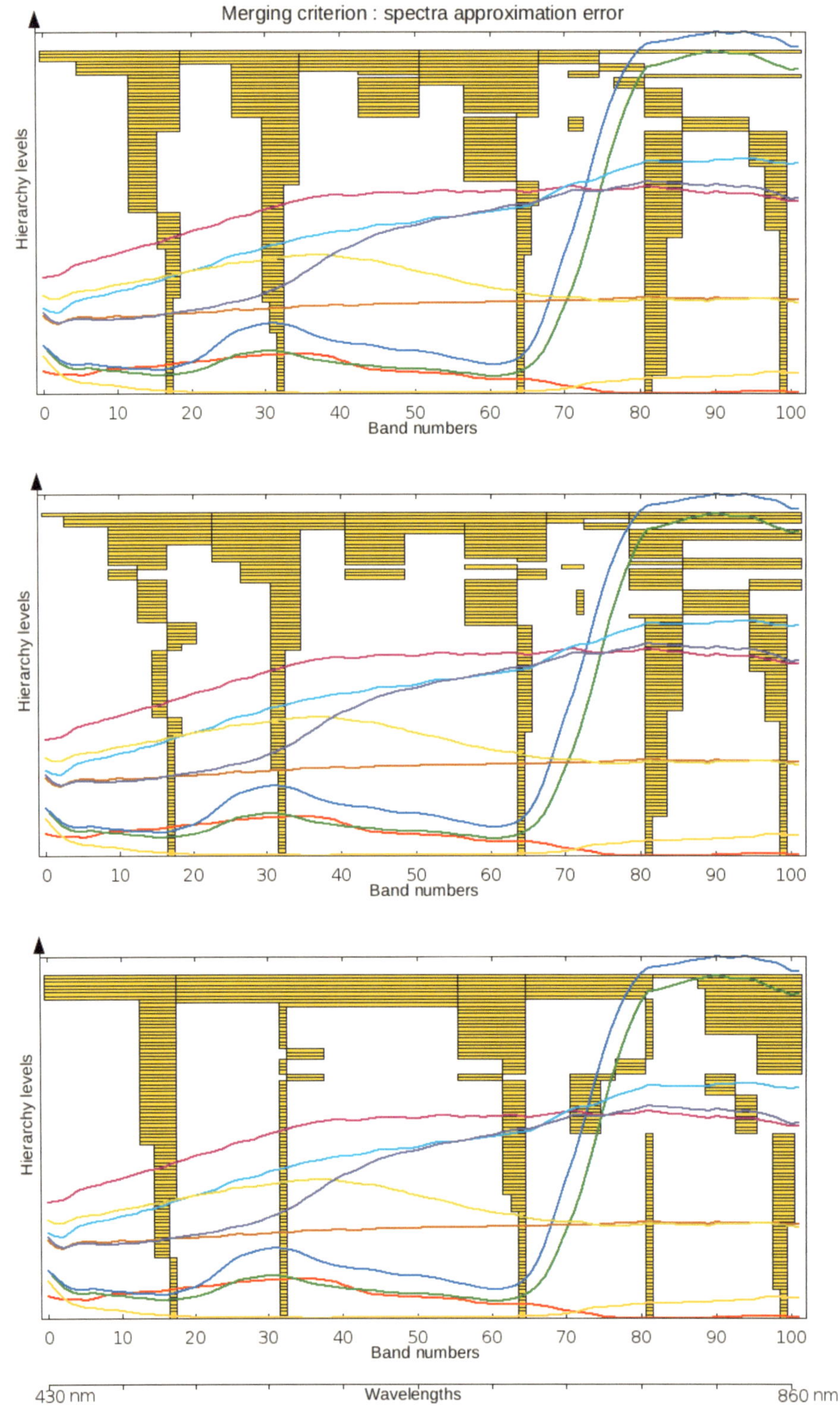

Figure 15.
Pavia data set: selected bands at the different levels of the hierarchy using the greedy approach for hierarchies of merged bands obtained using different band merging criteria: spectra piece-wise approximation error (top), between band correlation (middle) and class separability (bottom); x-axis corresponds to the band numbers/wavelengths; y-axis corresponds to the level in the band merging hierarchy (bottom—finest level with original bands; and top—only a single merged band).

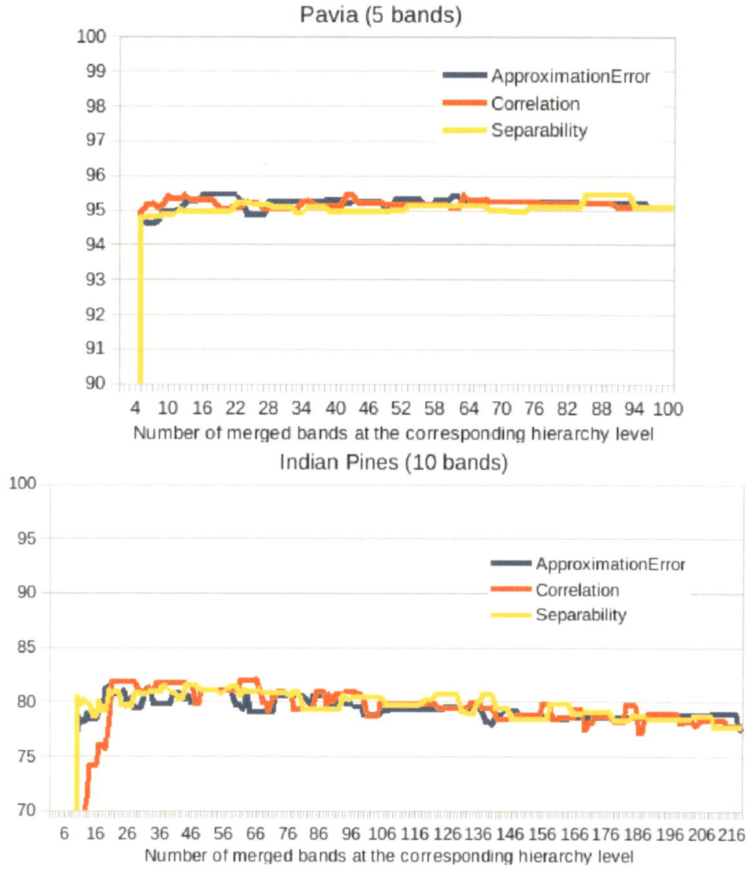

Figure 16.
Kappa (in %) reached for rbf SVM classification for merged band subsets selected at the different levels of the hierarchy for Pavia and Indian Pines data sets using the greedy FS algorithm (x-axis = number of merged bands in the spectral configuration corresponding to the hierarchy level).

The merged band subsets selected at the different levels of the hierarchy were evaluated according to a classification quality measure. As in the previous section, the Kappa coefficient reached by a rbf SVM was considered. Results for Pavia and Indian Pines data sets can be seen in **Figure 16**. At each level of the hierarchy, 5 bands were selected for Pavia, and 10 bands for Indian Pines. It can be seen that these accuracies remain very close to each other whatever the band merging criterion used, and no band merging criterion tends to really be better than the other ones. Results obtained using merged bands are generally better than using the original bands.

4.3.2 Taking into account the band merging hierarchy during selection

4.3.2.1 Proposed algorithm

The previous merged band selection approach is greedy and computing time expensive. So an adaptation of the SFFS heuristic was proposed to directly take into account the band merging hierarchy in the band selection process. As for the hierarchical band merging algorithm, a bottom-up approach was chosen. Contrary to the greedy approach, this one uses the band subset selected at the previous lower level when performing band selection at a new level of the hierarchy of merged bands. This algorithm is described below:

Let $S^{(l)} = \{S_i^{(l)}\}1 \le i \le p$ be the set of selected merged bands at level l of the hierarchy. (NB: the same number p of bands is selected at each level of the hierarchy.)

Initialization: standard SFFS band selection algorithm is applied to the base level $H^{(0)}$ of the hierarchy.

Iterations over the levels of the hierarchy:

Generate $S^{(l+1)}$ from $S^{(l)}$:

$$S^{(l+1)} \leftarrow \{L_l^{l+1}\left(S_i^{(l)}\right)\}_{1 \le i \le p}$$

Remove possible duplications from $S^{(l+1)}$.

if $\#S^{(l+1)} < p$,

find

$$s = argmax_{b \in H^{(l+1)}\ S^{(l+1)}} J\left(S^{(l+1)} \cup b\right)$$

$$S^{(l+1)} \leftarrow \{S^{(l+1)}; s\}$$

endif

Question $S^{(l+1)}$: find band $s \in S^{(l+1)}$ such that $S^{(l+1)} \backslash \{s\}$ maximizes FS score, i.e. $s = argmax_{z \in S^{(l+1)}} J\left(S^{(l+1)} \backslash \{s\}\right)$.

$$S^{(l+1)} \leftarrow S^{(l+1)} \backslash \{s\}$$

Then apply classic SFFS algorithm until $\#S^{(l+1)} = p$.

4.3.2.2 Results

Obtained results on Pavia scene for the band merging criterion 'spectra piece-wise approximation error' are presented in **Figure 17**: five merged bands were selected at each level of the hierarchy, starting from an initial solution obtained at the bottom level of the hierarchy.

As for previous experiments, obtained results were evaluated both for Pavia (5 selected bands) and Indian Pines (10 selected bands) data sets. Kappa reached for rbf SVM classification for merged band subsets selected at the different levels of the hierarchy (built for band merging criterion 'spectra piece-wise approximation

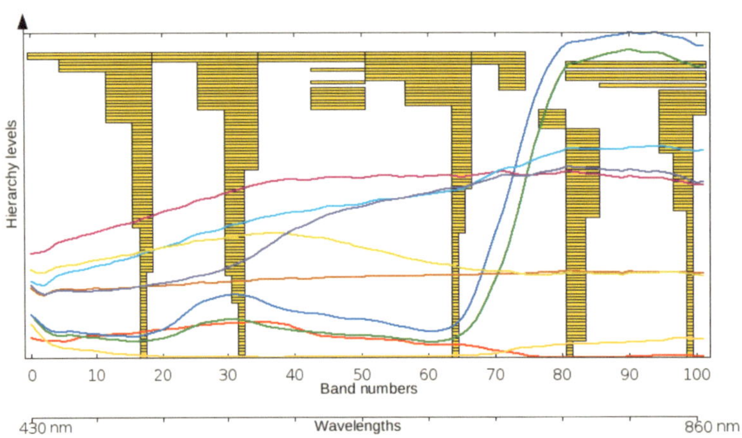

Figure 17.
Pavia data set: Selected bands at the different levels of the hierarchy using the proposed hierarchy aware algorithm for a hierarchy of merged bands obtained using spectra piece-wise approximation error band merging criteria.

error') can be seen both for the greedy FS algorithm and for the hierarchy aware one in **Figure 18**: obtained results remain very close, whatever the optimization algorithm.

Both algorithms lead to equivalent results considering classification performance (see **Table 4**), while the proposed hierarchy aware algorithm is really faster.

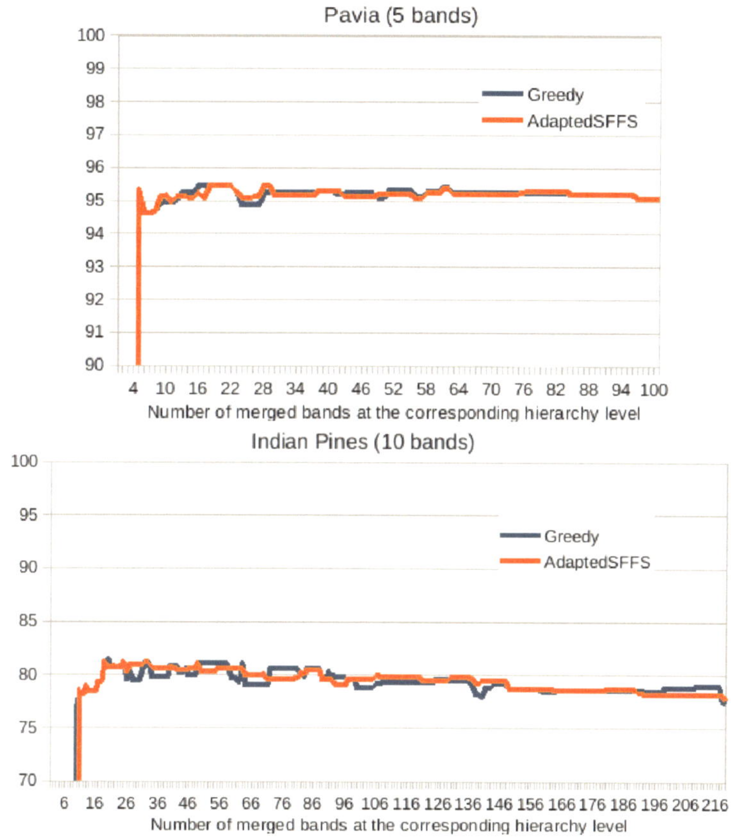

Figure 18.
Kappa (in %) reached for rbf SVM classification for merged band subsets selected at the different levels of the hierarchy (built for band merging criterion 'spectra piece-wise approximation error') for Pavia and Indian Pines data sets, using the hierarchy aware band selection algorithm.

	Original bands	Greedy SFFS	Adapted SFFS
Pavia (5 bands)			
Kappa (%)	95.05	95.45	95.44
Computing times	2min	1h10min	9min
Indian Pines (10 bands)			
Kappa (%)	77.69	81.41	81.21
Computing times	4min	7h	40min

Table 4.
Computing times and best kappa coefficients reached on Pavia (for a 5-band subset) and Indian Pines (for a 10-band subset) data sets for band merging criterion 'spectra piece-wise approximation error'.

5. Conclusion

Hyperspectral imagery consists of hundreds of contiguous spectral bands, but only a subset of well-chosen bands is generally sufficient for a specific classification problem. So it is possible to design superspectral sensors dedicated to specific land cover classification tasks. This chapter presented a spectral optimization strategy to identify the most relevant spectral band subset for such sensor, optimizing both band position and width. Spectral optimization involves a band subset relevance score as well as a method to optimize it.

This study first focused on the definition of this relevance score. Several filter, wrapper and embedded scores compatible with generic optimization heuristics were compared, and both their classification performance and selection stability were considered for band selection problem. At the end, most of them brought good results. Jeffries-Matusita distance score tended to lead to slightly better quantitative classification results than the best wrapper scores but also being less stable. Wrapper scores taking into account classification confidence performed better than classic wrapper scores expressed as a simple classification "hard label" error rate. For instance, a random forest confidence-based score was identified as one of the best criteria, considering both quantitative and qualitative analyses. As an intermediate result of this FS criteria comparison, a method to create band importance profiles according to the different criteria was proposed providing visual hints about the relevance of the different parts of the spectrum. Then the study focused on the optimization of bandwidth, which is important in a spectral sensor design context, as having wider bands is a way to limit signal noise while having too wide bands can also lead to a loss a useful information. A strategy consisting in building a hierarchy of groups of adjacent bands before applying band selection at the different levels of this hierarchy using an adaptation of an incremental algorithm for this problem. This band grouping strategy enabled to limit the problem's combinatory while considering relevant band subsets composed of spectral bands with different spectral widths. It was also a way to consider several possible solutions and evaluate their impact.

To conclude, algorithms proposed in this study were applied to design a sensor dedicated to classify urban materials [36, 72].

Author details

Arnaud Le Bris[1*†], Nesrine Chehata[1,2†], Xavier Briottet[3†] and Nicolas Paparoditis[1]

1 Université Paris-Est, LASTIG STRUDEL, IGN, ENSG, Saint-Mande, France

2 EA G&E Bordeaux INP, Bordeaux Montaigne University, Pessac, France

3 ONERA—The French Aerospace Lab, Toulouse, France

*Address all correspondence to: arnaud.le-bris@ign.fr

† These authors contributed equally.

References

[1] Hughes G. On the mean accuracy of statistical pattern recognizers. IEEE Transactions on Information Theory. 1968;**14**(1):55-63

[2] Camps-Valls G, Bruzzone L. Kernel-based methods for hyperspectral image classification. IEEE Transactions on Geoscience and Remote Sensing. 2005; **43**(6):1351-1362

[3] Melgani F, Bruzzone L. Classification of hyperspectral remote sensing images with support vector machines. IEEE Transactions on Geoscience and Remote Sensing. 2004;**42**(8):1778-1790

[4] Pal M, Foody G. Feature selection for classification of hyperspectral data by svm. IEEE Transactions on Geoscience and Remote Sensing. 2010;**48**(5): 2297-2307

[5] Demir B, Celebi A, Ertürk S. A low-complexity approach for the color display of hyperspectral remote-sensing images using one-bit-transform-based band selection. IEEE Transactions on Geoscience and Remote Sensing. 2009; **47**(1):97-105

[6] Le Moan S, Mansouri A, Voisin Y, Hardeberg J. A constrained band selection method based on information measures for spectral image color visualization. IEEE Transactions on Geoscience and Remote Sensing. 2011; **49**(12):5104-5115

[7] Clark RN, Swayze GA, Livo KE, Kokaly RF, Sutley SJ, Dalton JB, et al. Imaging spectroscopy: Earth and planetary remote sensing with the usgs tetracorder and expert systems. Journal of Geophysical Research. 2003;**108** (E12):5–1-5–44

[8] Bassani C, Cavalli R, Cavalcante F, Cuomo V, Palombo A, Pascucci S, et al. Deterioration status of asbestos-cement roofing sheets assessed by analyzing hyperspectral data. Remote Sensing of Environment. 2007;**109**:361-378

[9] Mohammadi M. Road classification and condition determination using hyperspectral imagery. The International Archives of the Photogrammetry, Remote Sensing and Spatial Information Sciences. 2012;**XXXIX-B7**

[10] Heiden U, Segl K, Roessner S, Kaufmann H. Determination of robust spectral features for identification of urban surface materials in hyperspectral remote sensing data. Remote Sensing of Environment. 2007;**111**(4):537-552

[11] Kira K, Rendell L. A practical approach to feature selection. In: Proceedings of the 9th International Workshop on Machine Learning. 1992. pp. 249-256

[12] Kononenko I, Simec E, Robnik-Sikonja M. Overcoming the myopia of inductive learning algorithms with relieff. Applied Intelligence. 1997;**7**(1): 39-55

[13] Guyon I, Weston J, Barnhill S, Vapnik V. Gene selection for cancer classification using support vector machines. Machine Learning. 2002;**46**: 289-422

[14] Estévez PA, Tesmer M, Perez CA, Zurada JM. Normalized mutual information feature selection. IEEE Transactions on Neural Networks. 2009; **20**(2):189-201

[15] Li S, Wu H, Wan D, Zhu J. An effective feature selection method for hyperspectral image classification based on genetic algorithm and support vector machine. Knowledge-Based Systems. 2011;**24**:40-48

[16] Du Q, Yang H. Similarity-based unsupervised band selection for

hyperspectral image analysis. IEEE Geoscience and Remote Sensing Letters. 2008;**5**(4):564-568

[17] Guo B, Damper R, Gunn S, Nelson J. A fast separability-based feature-selection method for high-dimensional remotely sensed image classification. Pattern Recognition. 2008;**41**:1653-1662

[18] Tibshirani R. Regression shrinkage and selection via the lasso. Journal of the Royal Statistical Society, Series B. 1996; **58**:267-288

[19] Breiman L, Friedman J, Stone C, Olshen R. Classification and Regression Trees. Boca Raton: CRC Press; 1984

[20] Campedel M, Maître H, Moulines E. Indexation Des Images Satellitaires—Comparaison et évaluation des caractéristiques pour la Classification. Télécom Paris: Tech. rep; 2004

[21] Díaz-Uriarte R, De Andres SA. Gene selection and classification of microarray data using random forest. BMC Bioinformatics. 2006;**7**(3):1-13

[22] Martínez-Usó A, Pla F, Martínez Sotoca J, García-Sevilla P. Clustering-based hyperspectral band selection using information measures. IEEE Transactions on Geoscience and Remote Sensing. 2007;**45**(12):4158-4171

[23] Tuia D, Volpi M, Dalla Mura M, Rakotomamonjy A, Flamary R. Automatic feature learning for spatio-spectral image classification with sparse SVM. IEEE Transactions on Geoscience and Remote Sensing. 2014;**52**(10): 6062-6074

[24] Serpico SB, Bruzzone L. A new search algorithm for feature selection in hyperspectral remote sensing images. IEEE Transactions on Geoscience and Remote Sensing. 2001;**39**:1360-1367

[25] Chang C-I, Wang S. Constrained band selection for hyperspectral imagery. IEEE Transactions on Geoscience and Remote Sensing. 2006; **44**(6):1575-1585

[26] Minet J, Taboury J, Pealat M, Roux N, Lonnoy J, Ferrec,Y. Adaptive band selection snapshot multispectral imaging in the vis/nir domain. In: Proceedings of SPIE the International Society for Optical Engineering; 2010. 7835; p. 10

[27] Le Bris, A, Chehata N, Briottet X, Paparoditis N. Identify important spectrum bands for classification using importances of wrapper selection applied to hyperspectral data. In: Proc. of the 2014 International Workshop on Computational Intelligence for Multimedia Understanding (IWCIM'14); 2014

[28] Fauvel M. Spectral and spatial methods for the classification of urban remote sensing data [Ph.D. thesis]. Institut National Polytechnique de Grenoble; 2007

[29] Hall MA, Holmes G. Benchmarking attribute selection techniques for discrete class data mining. IEEE Transactions on Knowledge and Data Engineering. 2003; **15**(6):1437-1447

[30] De Backer S, Kempeneers P, Debruyn W, Scheunders P. A band selection technique for spectral classification. IEEE Geoscience and Remote Sensing Letters. 2005;**2**(3): 319-323

[31] Ma S, Huang J. Penalized feature selection and classification in bioinformatics. Briefings in Bioinformatics. 2008;**8**(5):392-403

[32] Tuia D, Camps-Valls G, Matasci G, Kanevski M. Learning relevant image features with multiple-kernel classification. IEEE Transactions on Geoscience and Remote Sensing. 2010; **48**(10):3780-3791

[33] Pal M. Random forest classifier for remote sensing classification. International Journal of Remote Sensing. 2005;**26**(1):217-222

[34] Cariou C, Chehdi K, Le Moan S. Bandclust: An unsupervised band reduction method for hyperspectral remote sensing. IEEE Geoscience and Remote Sensing Letters. 2011;**8**(3): 565-569

[35] Bruzzone L, Serpico SB. A technique for feature selection in multiclass problem. International Journal of Remote Sensing. 2000;**21**(3): 549-563

[36] Le Bris A, Chehata N, Briottet X, Paparoditis N. Spectral band selection for urban material classification using hyperspectral libraries. ISPRS Annals. 2016;**3**(7):33-40

[37] Fröhlich H, Chapelle O, Schölkopf B. Feature selection for support vector machines by means of genetic algorithms. In: Proc. of the 15th IEEE International Conference on Tools with Artificial Intelligence; 2003. pp. 142-148

[38] Prasad S, Bruce LM. Decision fusion with confidence-based weight assignment for hyperspectral target recognition. IEEE Transactions on Geoscience and Remote Sensing. 2008; **46**(5):1448-1456

[39] Pal M. Margin-based feature selection for hyperspectral data. International Journal of Applied Earth Observation and Geoinformation. 2009; **11**:121-220

[40] Chang Y-L, Chen K-S, Huang B, Chang W-Y, Benediktsson J, Chang L. A parallel simulated annealing approach to band selection for high-dimensional remote sensing images. IEEE Journal of Selected Topics in Applied Earth Observations and Remote Sensing. 2011; **4**(3):579-590

[41] Breiman L. Random forests. Machine Learning. 2001;**45**(1):5-32

[42] Chang C. An information-theoretic approach to spectral variability, similarity and discrimination for hyperspectral image analysis. IEEE Transactions on Information Theory. 2000;**46**(15):1927-1932

[43] Zare A, Gader P. Hyperspectral band selection and endmember detection using sparsity promoting priors. IEEE Geoscience and Remote Sensing Letters. 2007;**5**(2):256-260

[44] Pudil P, Novovicova J, Kittler J. Floating search methods in feature selection. Pattern Recognition Letters. 1994;**15**:1119-1125

[45] Herold M, Gardner ME, Roberts DA. Spectral resolution requirements for mapping urban areas. IEEE Transactions on Geoscience and Remote Sensing. 2003;**41**(9): 1907-1919

[46] Kandasamy S, Tavin F, Minghelli-Roman A, Mathieu S, Weidong L, Baret F. et al. Optimization of image parameters using a hyperspectral library application to soil identification and moisture estimation. In: Geoscience and Remote Sensing Symposium,2009 IEEE International, IGARSS 2009; 2009. Vol. 3; pp. III-141-III-144

[47] Le Bris A, Chehata N, Briottet X, Paparoditis N. A random forest class memberships based wrapper band selection criterion: application to hyperspectral. In: Proc. of the 2015 IEEE International Geoscience and Remote Sensing Symposium (IGARSS'15); 2015

[48] Chang C-I, Du Q, Sun T-L, Althouse M. A joint band prioritization and band-decorrelation approach to band selection for hyperspectral image classification. IEEE Transactions on Geoscience and Remote Sensing. 1999; **37**(6):2631-2641

[49] Du H, Qi H, Wang X, Ramanath R, Snyder W. Band selection using independent component analysis for hyperspectral image processing. In: Proceedings of the 32nd Applied Imagery Pattern Recognition Workshop; 2003. pp. 93-98

[50] Hasanlou M, Samadzadegan F. ICA/PCA base genetically band selection for classification of hyperspectral images. In: Proc. of the Asian Conference on Remote Sensing (ACRS); 2010

[51] Kandasamy S, Minghelli-Roman A, Tavin F, Mathieu S, Baret F, Gouton P. Optimal band selection for future satellite sensor dedicated to soil science. In: Hyperspectral Image and Signal Processing: Evolution in Remote Sensing, 2009. WHISPERS '09. First Workshop; 2009. pp. 1-4

[52] Serpico SB, Moser G. Extraction of spectral channels from hyperspectral images for classification purposes. IEEE Transactions on Geoscience and Remote Sensing. 2007;**45**(2):484-495

[53] Yang H, Du Q, Chen G. Particle swarm optimization-based hyperspectral dimensionality reduction for urban land cover classification. IEEE Journal of Selected Topics in Applied Earth Observations and Remote Sensing. 2012;**5**(2):544-554

[54] Cang S, Hongnian Y. Mutual information based input feature selection for classification problems. Decision Support Systems. 2012;**54**: 691-698

[55] Battiti R. Using mutual information for selecting features in supervised neural net learning. IEEE Transactions on Neural Networks. 1994; **5**(4):537-550

[56] Peng H, Long F, Ding C. Feature selection based on mutual information: Criteria of max-dependency, max-relevance, and min-redundancy. IEEE Transactions on Pattern Analysis and Machine Intelligence. 2005;**27**(8): 1226-1238

[57] Sotoca J, Filiberto P. Supervised feature selection by clustering using conditional mutual information-based distances. Pattern Recognition. 2010; **43**(6):2068-2081

[58] Yang H, Zhang S, Deng K, Du P. Research into a feature selection method for hyperspectral imagery using PSO and SVM. Journal of China University of Mining and Technology. 2007;**17**(4): 473-478

[59] Zhuo L, Zheng J, Wang F, Li X, Bin A, Qian J. A genetic algorithm based wrapper feature selection method for classification of hyperspectral images using support vector machine. The International Archives of the Photogrammetry, Remote Sensing and Spatial Information Sciences. 2008;**37** (B7):397-402

[60] Fauvel M, Zullo A, Ferraty F. Nonlinear parsimonious feature selection for the classification of hyperspectral images. In: Proc. of the 6th Workshop on Hyperspectral Image and Signal Processing: Evolution in Remote Sensing (WHISPERS'14); 2014

[61] Zhang L, Zhong Y, Huang B, Gong J, Li P. Dimensionality reduction based on clonal selection for hyperspectral imagery. IEEE Transactions on Geoscience and Remote Sensing. 2007; **45**(12):4172-4186

[62] Tang J, Alelyani S, Liu H. Feature selection for classification: A review. In: Data classification: Algorithms and applications. Chapman and Hall/CRC data mining and knowledge discovery series. Boca Raton: CRC Press; 2014. pp. 37-64

[63] Tuia D, Courty N, Flamary R. A group-lasso active set strategy for

multiclass hyperspectral image classification. ISPRS Annals of the Photogrammetry, Remote Sensing and Spatial Information Sciences. 2014;**II-3**: 1-8

[64] Zhu J, Rosset S, Hastie T, Tibshirani R. 1-norm support vector machines. Advances in Neural Information Processing Systems. 2004; **16**(1):49-56

[65] Zhou S, Zhang J, Su B. Feature selection and classification based on ant colony algorithm for hyperspectral remote sensing images. In: Proc. of the 2nd International Congress on Image and Signal Processing (CISP'09); 2009. pp. 1-4

[66] Su H, Yang H, Du Q, Sheng Y. Semisupervised band clustering for dimensionality reduction of hyperspectral imagery. IEEE Geoscience and Remote Sensing Letters. 2011;**8**(6): 1135-1139

[67] Boser E, Guyon I, Vapnik V. A training algorithm for optimal margin classifiers. In: Fifth annual workshop on computational learning. Theory. 1992;**5**: 144-152

[68] Bigdeli B, Samadzadegan F, Reinartz P. Band grouping versus band clustering in svm ensemble classification of hyperspectral imagery. Photogrammetric Engineering and Remote Sensing. 2013; **79**(6):523-533

[69] Jensen A-C, Solberg A-S. Fast hyperspectral feature reduction using piecewise constant function approximations. IEEE Geoscience and Remote Sensing Letters. 2007;**4**(4): 547-551

[70] Wiersma D, Landgrebe D. Analytical design of multispectral sensors. IEEE Transactions on Geoscience and Remote Sensing. 1980; **GE-18**(2):180-189

[71] Adeline K, Gomez C, Gorretta N, Roger J. Sensitivity of soil property prediction obtained from VNIR/SWIR data to spectral configurations. In: Proc. of the 4th International Symposium on Recent Advances in Quantitative Remote Sensing: RAQRS'IV; 2014

[72] Le Bris A, Chehata N, Briottet X, Paparoditis N. Hierarchically exploring the width of spectral bands for urban material classification. In: Proc. of JURSE; 2017

Chapter 6

Clustering Techniques for Land Use Land Cover Classification of Remotely Sensed Images

Debasish Chakraborty

Abstract

Image processing is growing fast and persistently. The idea of remotely sensed image clustering is to categorize the image into meaningful land use land cover classes with respect to a particular application. Image clustering is a technique to group an image into units or categories that are homogeneous with respect to one or more characteristics. There are many algorithms and techniques that have been developed to solve image clustering problems, though, none of the method is a general solution. This chapter will highlight the various clustering techniques that bring together the current development on clustering and explores the potentiality of those techniques in extracting earth surface features information from high spatial resolution remotely sensed imageries. It also will provide an insight about the existing mathematical methods and its application to image clustering. Special emphasis will be given on Hölder exponent (HE) and Variance (VAR). HE and VAR are well-established techniques for texture analysis. This chapter will highlight about the Hölder exponent and variance-based clustering method for classifying land use/land cover in high spatial resolution remotely sensed images.

Keywords: remote sensing image, clustering, classification, land use, land cover, features, extraction

1. Introduction

High spatial resolution remotely sensed imagery helps to obtain quality and detailed information about the earth's surface features in conjunction with their geographical associations. The internal changeability within the identical land-use land-cover units augments with the rise of resolution. The augmented changeability diminishes the statistical distinguishability of land-use/land-cover classes in the spectral data space. This reduced distinguishability tends to decrease the accuracies of pixel-based clustering algorithms such as Fuzzy C Means [1], minimum distance classifiers [2] and K-Means [3]. These pixel-based clustering techniques assign a pixel to a region according to the similarities of spectral signature. It considers only one pixel at a time [4]. Spectral signatures are the specific combination of emitted, reflected or absorbed electromagnetic (EM) radiation at varying wavelengths which can uniquely identify an object [4].

Compared to IRS-1A/1B sensors, the spectral resolution of high spatial resolution images is normally relatively poor. Spectral resolution describes a sensor's ability to identify fine intervals of wavelength. The better the spectral resolution, the finer the channel or band width. Therefore, between spatial and spectral resolution, there is a trade-off. It is mainly true for panchromatic (PAN) images of high spatial resolution, namely CARTOSAT-II 1m and IKONOS 1m. There is a need to consider the spatial relationships between pixel values, also known as the 'texture' of the scene objects to classify high-resolution (HR) images owing to the wide difference in the spatial structure in these images. Consequently, multiple texture-based clustering technique namely GLCM [5–8], Markov random field (MRF) model [5], Gray scale rotation invariant [9] were evolved for clustering remote sensing images having high spatial resolution. Nevertheless, above mentioned methods are appropriate in textured area of HR images. A region is called textured; where the intensity dissimilarity within adjacent pixels is substantial. A region is said to be non-textured, where the intensity dissimilarity among adjacent pixels is insignificant [10, 11]. But texture-based classification techniques failed in non-textured region of high spatial resolution image as much variation is not found in the spatial pattern of those regions of the image [12]. Thus, we can infer from earlier studies that classification of high spatial resolution imageries either by pixel or texture-based algorithm may not yield desired results.

Some more techniques namely watershed approach [13, 14], region-growing approach [4, 15], mean shift approach [16, 17], region merging approach [18] etc. are in use for clustering high spatial resolution remote sensing images. Application of these approaches for clustering of images either leads to under-segmentation or over-segmentation [19, 20]. Structural image indexing approach [21], semi-supervised feature learning approach [22] and multi-scale manner using SVM approach [23] are also found fairly suitable in clustering high resolution images. The imagery of higher resolution includes textured and non-textured areas. Hence, pixel or texture-based algorithm for clustering of high-resolution imagery does not produce expected results. This type of high-resolution imagery clustering research is in the trend. Multi-circular local binary pattern and variance-based method [10] were used separately to cluster high resolution image having textured and non-textured regions. The Multi circular local binary pattern operator has been used here for measuring the spatial structure of the image. But, disadvantage in this strategy is that multi-circular local binary pattern operator is susceptible to noise as it exactly sees the value of the moving window's central pixel as a limit for computing the spatial structure around the central pixel.

In last one decade the Hölder exponent (HE) has been used for calculating spatial structure of the images [24–26]. It is also being used for clustering high-resolution images [12]. HE gives an evidence of the spatial structure of the image and is not much influenced by the noise. In addition, spatial structure, contrast of the local image holds considerable property for calculating the texture around the pixel. In this research, high-resolution picture textured and non-textured region is originally segmented using HE and VAR-based method and subsequently separately clustered and non-textured areas. VAR is used to calculate the contrast around the pixel. The suggested method is applied with a 1 m spatial resolution on high resolution IKONOS PAN images.

2. Methods

The suggested high-resolution image 'P' clustering technique has three main steps: (i) image transformation, (ii) segmentation and extraction, and (iii) clustering. Initially, every pixel of the image is converted into a degree of texture or non-texture around the pixel. In the second step, using segmented image mask, the transformed image is segmented and non-textured and textured regions are extracted from the initial image. Finally, the two areas obtained are separately clustered.

2.1 Transformation of image

The Hölder Exponent (HE) and VAR are jointly used to convert the image for computing the texture. The HE calculates each pixel of P's spatial structure. Besides spatial structure, local image contrast also grasps important property for computing the texture around the pixel. In this research, therefore, VAR is used to calculate the contrast around the pixel.

2.1.1 Hölder exponent

Hölder exponent has been used for investigating the texture in high-resolution images [12]. It measures the irregularity in the vicinity. Supremacy of applying Hölder Exponent in HR images are that (i) it can be used as an instrument to calculate each pixel of the image's spatial structure, (ii) no previous data on the pixel intensity is required and (iii) is not very sensitive to noise [12].

Definition of HE [27]: Let μ be a measure on a set Ω as well as for all $x \in \Omega$, $\ni \alpha(x)$, such that $\mu\ (B_r(x)) \sim r^{\alpha}$, for small r. Here $B_r(x)$ is circle (2D) of radius i centered on x. Then $\alpha\ (x)$ is called the HE on x.

A sequence of 15 values of radius r (i.e. 1, $\sqrt{2}$, $\sqrt{5}$, 3, $\sqrt{13}$, $3\sqrt{2}$, 5, $\sqrt{29}$, $2\sqrt{10}$, $3\sqrt{5}$, 7, $\sqrt{61}$, $6\sqrt{2}$, $\sqrt{85}$, $7\sqrt{2}$) centered on x are used as a scale parameter for calculating HE value around each pixel x in the image [12] and the total number (N) of intersected pixels by the perimeter of series of circles of radius r is considered as a scale parameter for computing VAR value around x [12]. N is computed using Eq. (1).

$$N = \sum_{r=1}^{t} m_r \tag{1}$$

where t is the total number of identified circles, m_r is the number of intersected pixels on the perimeter of the radius r circle.

2.1.2 VAR (σ^2) for contrast measurement around each pixel of the image

To get the contrast value of (x, y), the neighbor's σ^2 of each pixel (x, y) is calculated over the entire image. Using Eq. (2), the σ^2 (x, y) is realized

$$\sigma^2(x,y) = \frac{\sum_{r=1}^{t}\sum_{j=1}^{m_r}\left(a_{rj} - \mu\right)^2}{N} \tag{2}$$

where a_{rj} is the intensity value of pixel (r,j),

$$\mu = \frac{\sum_{r=1}^{t}\sum_{j=1}^{m_r} a_{rj}}{N}$$

Thus obtained $\alpha(x,y)$ and σ^2 (x,y) for each P(x,y). Afterward, these values are used in Eq. (3) to obtain the corresponding pixel value (x,y) in the transformed image T. Each pixel (x,y) of T signifies the degree of texture around that pixel.

$$T(x,y) = \frac{\alpha(x,y) + \sigma^2(x,y)}{2} \tag{3}$$

2.2 Image segmentation and extraction

The image 'T' is segmented into textured and non-textured regions based on a threshold value 'δ'. The pixel value in T below the 'δ' is considered to be a non-textured region, whereas greater than or equal to 'δ' is considered to be the textured region in the segmented image. Pixels are labeled as zero in non-textured areas, whereas pixels are marked as one in textured areas in the segmented image mask and depicted as follows:

$$\Gamma(\mathrm{x}, \mathrm{y}) = \begin{cases} 1, T(x, y) \geq \delta \\ 0, T(x, y) < \delta \end{cases} \tag{4}$$

where T(x,y) and Γ(x,y) represents the pixel value in (x,y) position of the two dimensional transformed image and segmented image respectively and δ represents the threshold value. The δ is calculated by using Eq. (5).

$$\delta = \mathrm{T_{min}} + \frac{\mathrm{T_{max}} - \mathrm{T_{min}}}{\mathrm{K}} \tag{5}$$

where T_{min} and T_{max} represents minimum and maximum pixel gray value in T respectively and K is user defined value.

IKONOS PAN sensor image of size 256 × 256 pixels (shown in **Figure 1a**) is used to achieve the optimum K. The suggested clustering method is also implemented for distinct K values on this image.

The segmented image is subsequently used to obtain the textured and non-textured region from the initial image P. This process's mathematical representation is shown as follows:

$$\mathrm{R_1}(\mathrm{x}, \mathrm{y}) = \begin{cases} \mathrm{P}(\mathrm{x}, \mathrm{y}), & \Gamma(\mathrm{x}, \mathrm{y}) = 0 \\ 0, & \Gamma(\mathrm{x}, \mathrm{y}) \neq 0 \end{cases} \tag{6}$$

$$\mathrm{R_2}(\mathrm{x}, \mathrm{y}) = \begin{cases} \mathrm{P}(\mathrm{x}, \mathrm{y}), & \Gamma(\mathrm{x}, \mathrm{y}) = 1 \\ 0, & \Gamma(\mathrm{x}, \mathrm{y}) \neq 1 \end{cases} \tag{7}$$

where P, Γ, R_1 and R_2 indicates original image, segmented image, extracted non-textured region from original image P and extracted textured region from original image P respectively.

2.3 Clustering

Initially, a threshold is used to segment the transformed image into textured and non-textured region. Afterward, the original image is extracted into textured and non-textured regions using the segmented image mask and clustered independently. The extracted textured region (R_2) is clustered by means of ISODATA clustering algorithm [28] considering HE, VAR and intensity values of individual pixel of textured area. The clustering algorithm of ISODATA is less computational, easy and non-supervisory. Whereas the non-textured area (R1) of the image is categorized using the clustering algorithm of ISODATA. In the event of non-textured region, the individual pixel HE and VAR value is not regarded for classification as there is no important variation in texture between classes. The classified outputs of the non-textured and textured region are subsequently produced separately and mixed together to obtain the final classified image.

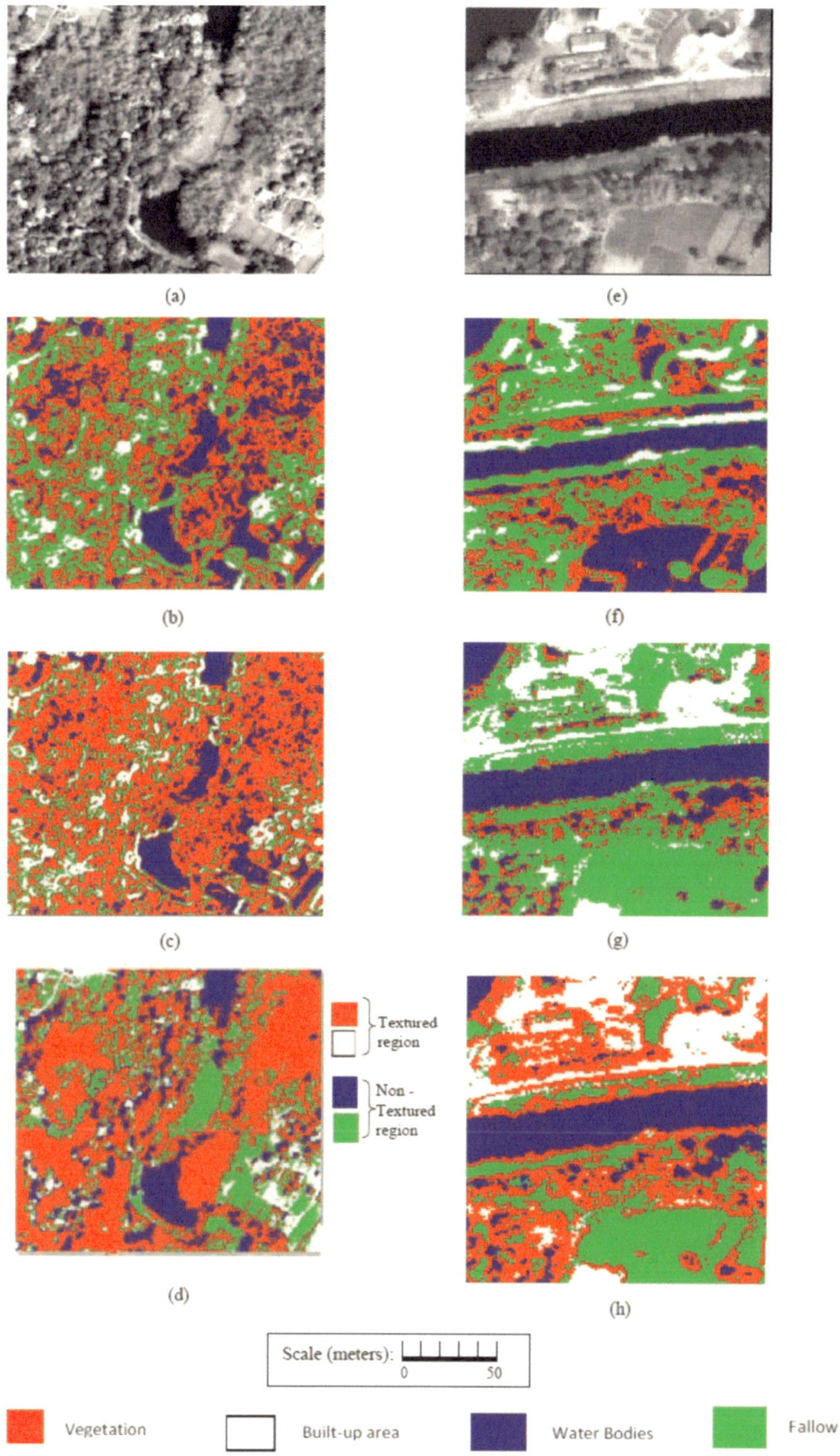

Figure 1.
(a) IKONOS image showing vegetation, built-up area, fallow and water body categories, (b) classified image obtained by applying "HE-VAR and PAN" based method on ***Figure 1a****, (c) classified image obtained by applying "MCLBP and VAR" based method on* ***Figure 1a****, (d) classified image obtained by applying "proposed classification method" on* ***Figure 1a****, (e) IKONOS image showing fallow, water bodies, vegetation and built-up area categories, (f) classified image obtained by applying "HE-VAR and PAN" based method on* ***Figure 1e****, (g) classified images obtained by applying "MCLBP and VAR" based method on* ***Figure 1e****, (h) Classified images obtained by applying "proposed classification method" on* ***Figure 1e****.*

This research uses "HE-VAR and PAN" and "MCLBP and VAR" based clustering technique to show the power of the suggested clustering technique. The technique based on "HE-VAR and PAN" clusters the entire image using the HE, VAR and intensity of each pixel of the IKONOS PAN image. The suggested technique of clustering is then contrasted with the outcomes of the clustering method based on "HE-VAR and PAN" and "MCLBP and VAR" to demonstrate the strength of the suggested technique of clustering.

3. Results and discussion

The projected clustering method imagines threshold δ to get the segmented image mask from the transformed image. The threshold is computed using a constant 'K'. In this study, proposed clustering procedure is implemented on IKONOS PAN image with spatial resolution 1 m for 'K' values between 3 and 7 and subsequently, classification rate is measured for these 'K' values using the ground truth data. The classification accuracy with different 'K' is shown in **Figure 2**. The 'K' affects the accuracy in classifying High spatial resolution images considerably as shown in **Figure 2**. For computing texture, a suitable choice of 'K' is important. In this study, superlative performance in high-resolution image classification was accomplished with K = 5. The optimum K is discovered based on **Figure 1a** and is also implemented in the classification of **Figure 1e** in addition to other images and found classification accuracy is more than 88%. Thus, from the present study, we can infer that the same K value is suitable for most images.

The Proposed clustering method, "MCLBP and VAR" based method and "HE-VAR and PAN" based method were applied on two different 1 m PAN (IKONOS) images (size 256 × 256 pixels) covering (i) vegetation, (ii) built-up area, (iii) water bodies, and (iv) fallow (shown in **Figure 1a, e**). Texture is observable in in **Figure 1a, e**. The results of proposed method are then compared with the results obtained from the analysis based on "HE-VAR and PAN" and "MCLBP and VAR" respectively.

Figure 1f–h shows the classification outcomes of the methods "HE-VAR and PAN," "MCLBP and VAR" and "Proposed classification" after proceeding to the second IKONOS image respectively. **Figure 1b–d** shows the classification outcomes of the methods "HE-VAR and PAN," "MCLBP and VAR" and "Proposed classification" after proceeding to the first IKONOS image respectively. Classified images recognize varied features in **Figure 1b–d, f–h**. From the results, it is evident that the

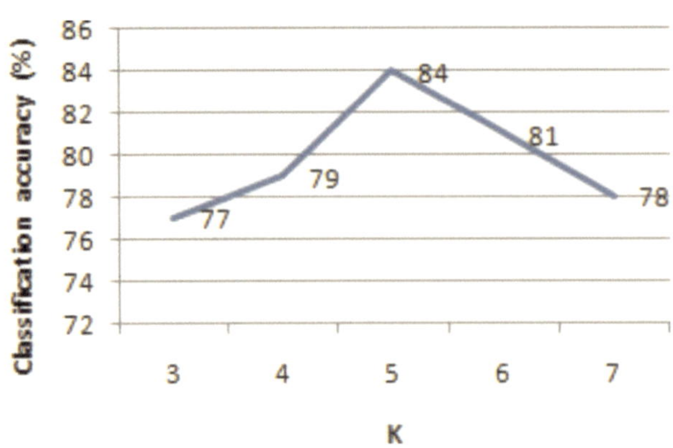

Figure 2.
Classification accuracy as a function of K.

method based on "MCLBP and VAR" gives less heterogeneous segments than the method based on "HE-VAR and PAN," while the method based on "Proposed classification method" provides more homogeneous segments with distinct classes than the method based on "MCLBP and VAR."

The ground truth data is collected using GPS equipment for the class vegetation, built-up area, fallow and water body of sample size of 656, 519, 577 and 462 square meters respectively. Afterward, ArcGIS software is used to transfer the ground truth data into vector data. Subsequently, by overlaying the ground truth information distinctly on the results acquired from both IKONOS images (**Figure 1a, e**) adopting methods such as "HE-VAR and PAN," "MCLBP and VAR" and "Proposed clustering," the classification accuracies for each strategy are shown by confusion matrix. The confusion matrices (**Table 1**) calculated for **Figure 1b–d** showed that the precision of classification of vegetation, built-up area, fallow and water bodies is (73, 69, 59 and 87% respectively) based on the 'HE-VAR and PAN' technique and (79, 71, 68 and 89% respectively) based on the 'MCLBP and VAR 'technique, whereas (91, 86, 85 and 94% respectively) by the "Proposed clustering" method. **Table 2** demonstrates the confusion matrices calculated for **Figure 1f–h** showed that the precision of classification of vegetation, built-up area, fallow and water bodies is (73, 74, 66 and 88% respectively) based on the 'HE-VAR and PAN 'technique and (78, 76, 68 and 89% respectively) based on the 'MCLBP and VAR 'technique whereas (90, 87, 86 and 93% respectively) by the "Proposed clustering" method.

The categorized result for **Figure 1a, e** shows that the "HE-VAR and PAN" method under segment as a result (i) fallow assorted with water bodies shown in **Figure 1b, f**, (ii) built-up region assorted with fallow and vegetation shown in **Figure 1f**, (iii) vegetation assorted with water bodies shown in **Figure 1b, f**, (iv) fallow assorted with built-up region shown in **Figure 1b**. This incoherence decreases vegetation, fallow, water bodies and built-up area classification precision as shown in **Tables 1** and **2**. The technique based on "MCLBP and VAR" somehow overcomes these inconsistencies. It is discovered that, as shown in **Figure 1c, g**, the superposition of fallow, water body, vegetation region becomes less. In addition, decreased inconsistencies improve the accuracy of the classification of fallow, water body and vegetation regions (see **Tables 1** and **2**).

"HE-VAR and PAN" based method classifies water bodies and fallow areas as a single class (**Figure 1b, f**) since the texture patterns of these two areas does not show much difference in high resolution imageries as shown in **Figure 1a, e**. "MCLBP and VAR" based technique demonstrates improvement in classifying the fallow areas and water bodies which is observable in **Figure 1g**. But this method could not extract non-textured region appropriately form **Figure 1a** since MCLBP is sensitive to noise. Therefore "MCLBP and VAR" based method could not discriminate appropriately fallow areas and water bodies in **Figure 1a** as visible in **Figure 1c**. HE is not as much of sensitive to noise therefore the proposed technique partitions the image into textured and non-textured regions noticeably which in turn helps in classifying the fallow and water bodies as shown in **Figure 1d**.

The proposed clustering method is applied further on a 1 m PAN (IKONOS) image (**Figure 3a**) of (i) urban woodland, (ii) building, (iii) water bodies, and (iv) fallow to show the robustness and validity of the method in classifying land use area. The method satisfactorily discriminate urban woodland, building, fallow and water bodies as shown in **Figure 3b**. The algorithm also implemented on two extra 1 m PAN (IKONOS) images: (i) **Figure 4a** of fallow, vegetation, built-up area and bare land and (ii) **Figure 4c** of water, vegetation, fallow and built-up area. The findings (**Figure 4b, d**) show that vegetation, fallow, built-up region, bare soil and water bodies are satisfactorily discriminated against by the algorithm.

		Classes derived from satellites							
Classification method	**Grand observed class**	**Vegetation**	**Built-up area**	**Fallow**	**Water body**	Row total	**Classification accuracy (%)**	**C (%)**	**O (%)**
HE-VAR and PAN	Vegetation	251	35	12	7	305	72.75	27.24	15.65
	Built-up area	18	163	14	5	200	69.07	30.93	15.68
	Fallow	53	23	144	14	234	59.26	40.74	37.04
	Water body	23	15	73	177	288	87.19	12.81	54.67
	Column total	345	236	243	203	1027			
MCLBP and VAR	Vegetation	272	31	11	4	332	78.84	18.26	13.33
	Built-up area	17	168	7	3	205	71.25	28.81	11.41
	Fallow	43	21	166	15	223	68.33	31.69	32.51
	Water body	13	16	59	181	258	89.28	10.83	43.35
	Column total	345	236	243	203	1027			
Proposed method	Vegetation	313	17	7	3	340	90.85	9.27	7.82
	Built-up area	7	204	8	2	221	86.28	13.56	7.20
	Fallow	19	7	206	7	239	84.77	15.23	13.58
	Water body	6	8	22	191	227	94.11	5.91	17.73
	Column total	345	236	243	203	1027			

C: Commission error, O: Omission error.

Table 1.
The confusion matrices showing the classification accuracy obtained by applying "HE-VAR and PAN", "MCLBP and VAR" and "Proposed" methods separately on IKONOS image shown in ***Figure 1a****.*

		Classes derived from satellites							
Classification method	**Grand observed class**	**Vegetation**	**Built-up area**	**Fallow**	**Water body**	Row total	**Classification accuracy (%)**	**C (%)**	**O (%)**
HE-VAR and PAN	Vegetation	228	15	43	16	302	73.3	26.68	23.79
	Built-up area	32	209	14	5	260	73.7	26.14	18.02
	Fallow	27	27	221	11	286	66.3	33.83	19.46
	Water body	24	32	56	227	339	87.6	12.35	43.24
	Column total	311	283	334	259	1187			
MCLBP and VAR	Vegetation	241	13	38	9	301	77.6	22.51	20.58
	Built-up area	29	215	13	7	264	75.83	24.02	17.31
	Fallow	26	26	228	11	291	68.36	31.74	18.86
	Water body	15	29	55	232	331	89.4	10.42	38.22
	Column total	311	283	334	259	1187			
Proposed method	Vegetation	281	7	14	6	308	90.4	9.64	8.68
	Built-up area	3	244	9	5	261	86.7	13.78	6.0
	Fallow	15	15	287	7	324	85.8	14.07	11.07
	Water body	12	17	24	241	294	93.2	6.94	20.46
	Column total	311	283	334	259	1187			

C: Commission error, O: Omission error.

Table 2.
The confusion matrices showing the classification accuracy obtained by applying "HE-VAR and PAN", "MCLBP and VAR" and "Proposed" methods separately on IKONOS image shown in ***Figure 1e****.*

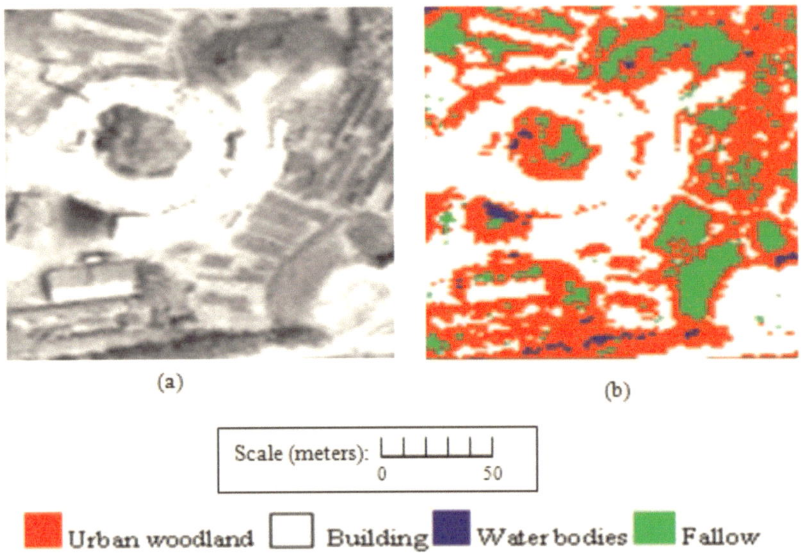

Figure 3.
*(a) IKONOS image showing urban woodland, building, water body and fallow categories, (b) classified image obtained by applying "proposed classification method" on **Figure 3a**.*

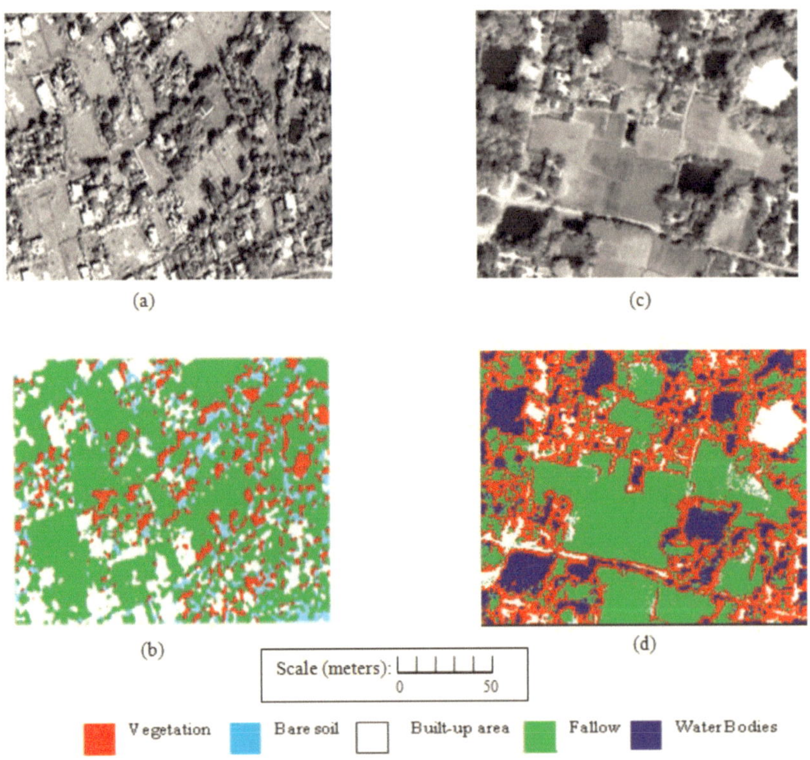

Figure 4.
*(a) IKONOS image showing fallow, built-up area, vegetation and bare soil categories, (b) classified image obtained by applying "proposed classification method" on **Figure 4a**, (c) IKONOS image showing vegetation, fallow, built-up area and water bodies categories, (d) classified image obtained by applying "proposed classification method" on **Figure 4c**.*

4. Conclusion

In the present study, the spatial structure of local image texture is computed using HE. The contrast around the pixel is measured using VAR. Afterward, the image is transformed using HE and VAR together for measuring the texture. A threshold δ is used to extract textured and non-textured region from the image. The classification algorithm ISODATA is used to classify the textured region taking into account HE, VAR and intensity values of the textured area's individual pixels. Whereas ISODATA clustering algorithm classifies the extracted non-textured region of the image. The HE and VAR value of individual pixels is not regarded for classification in the event of non-textured region. From the research outcomes, it is discovered that the suggested technique is helpful to extract earth surface characteristics from complicated remote sensing images that contain both textured and non-textured areas. Moreover, it can be considered as an intuitively appealing and unsupervised clustering algorithm for extracting features from remotely sensed images. As a result, the method is potentially useful to extract earth surface features by clustering high spatial resolution panchromatic images more efficiently.

Acknowledgements

The author sincerely thanks the Director, NRSC, Hyderabad, India and CGM, RCs, NRSC, Hyderabad, India for their support. The author is also grateful to the former GM, RRSC-East for support.

Author details

Debasish Chakraborty
Regional Remote Sensing Centre-East, National Remote Sensing Centre, ISRO, New Town, Kolkata, West Bengal, India

*Address all correspondence to: deba.isro@gmail.com

References

[1] Bezdek JC, Ehrlich R, Full W. FCM: The Fuzzy C-Means clustering algorithm. Computers and Geosciences. 1984;**10**:191-203

[2] Richards JA. Remote Sensing Digital Image Analysis: An Introduction. Berlin, Heidelberg: Springer-Verlag; 1995. pp. 265-290

[3] Hartigan JA, Wong MA. A K-means clustering algorithm. Applied Statistics. 1979;**28**(1):100-108

[4] Chakraborty D, Sen GK, Hazra S. Image Segmentation Techniques. Germany: LAP LAMBERT Academic Publishing; 2012. pp. 1-128

[5] Clausi DA, Yue B. Texture segmentation comparison using grey level co-occurrence probabilities and Markov random fields. In: Proceedings of the 17th International Conference on Pattern Recognition (ICPR'04); 2004. Available from: http://ieeexplore.ieee.org/stamp/

[6] Haralick RM, Shanmugan K, Dinstein I. Textural features for image classification. IEEE Transactions on Systems, Man, and Cybernetics. 1973;**6**: 610-621

[7] Tsai F, Chou MJ. Texture augmented analysis of high resolution satellite imagery in detecting invasive plant species. Journal of the Chinese Institute of Engineers. 2006;**29**(4):581-592

[8] Tsai F, Chou MJ, Wang HH. Texture analysis of high resolution satellite imagery for mapping invasive plants. In: International Geoscience and Remote Sensing Symposium; 2005. pp. 43024-43027

[9] Klemas VV. Remote sensing of coastal resources and environment. Environmental Research, Engineering and Management. 2009; **2**(48):11-18

[10] Chakraborty D, Chowdhary VM, Dutta D, Sharma JR. Classification of high spatial resolution image using multi circular local binary pattern and variance. International Journal of Electronics and Computer Engineering. 2013;**4**(6):1648-1653

[11] Weng Q. Remote sensing of impervious surfaces in the urban areas: Requirements, methods, and trends. Remote Sensing of Environment. 2012; **117**:34-49

[12] Chakraborty D, Sen GK, Hazra S. High-resolution satellite image segmentation using H¨older exponents. Journal of Earth System Science. 2009; **118**(5):609-617

[13] Mathivanan B, Selvarajan S. High spatial resolution remote sensing image segmentation using marker based watershed algorithm. Journal of Academia and Industrial Research. 2012; **1**(5):257-260

[14] Wang Z, Zhao S, Chen X. Watershed segmentation of high-resolution remotely sensed data. In: Proceedings of International Symposium on Remote Sensing & 20th Anniversary of the Korean Society of Remote Sensing; 2004. pp. 107-109

[15] Carleer AP, Debeir O, Wolff E. Assessment of very high spatial resolution satellite image segmentations. Photogrammetric Engineering and Remote Sensing. 2005;**71**(11):1285-1294

[16] Chakraborty D, Sen GK, Hazra S, Jeyaram A. Clustering for high resolution monochrome satellite image segmentation. International Journal of Geoinformatics. 2008;**4**(1):1-9

[17] Su T, Li H, Zhang S, Li Y. Image segmentation using mean shift for extracting croplands from high-resolution remote sensing imagery.

Remote Sensing Letters. 2015;**6**(12): 952-961

[18] Zhang X, Xiao P, Feng X, Wang J, Wang Z. Hybrid region merging method for segmentation of high-resolution remote sensing images. ISPRS Journal of Photogrammetry and Remote Sensing. 2014;**98**:19-28

[19] Chen B, Qiu F, Wu B, Du H. Image segmentation based on constrained spectral variance difference and edge penalty. Remote Sensing. 2015;7: 5980-6004

[20] Wang C, Shi AY, Wang X, Wu FM, Huang FC, Xu LZ. A novel multi-scale segmentation algorithm for high resolution remote sensing images based on wavelet transform and improved JSEG algorithm. Optik - International Journal for Light and Electron Optics. 2014;**125**(19):5588-5595

[21] Xia GS, Yang W, Delon J, Gousseau Y, Sun H, Maître H. Structural high-resolution satellite image indexing. In: ISPRS TC VII Symposium-100 Years ISPRS, Vol. 38; 2010. pp. 298-303

[22] Yang W, Yin X, Xia GS. Learning high-level features for satellite image classification with limited labeled samples. IEEE Transactions on Geoscience and Remote Sensing. 2015; **53**(8):4472-4482

[23] Huang X, Zhang L. An SVM ensemble approach combining spectral, structural, and semantic features for the classification of high-resolution remotely sensed imagery. IEEE Transactions on Geoscience and Remote Sensing. 2013;**51**(1):257-272

[24] Lucieer A, Stein A, Fisher P. Multivariate texture-based segmentation of remotely sensed imagery for extraction of objects and their uncertainty. International Journal of Remote Sensing. 2005;**26**(14): 2917-2936

[25] Malladi RK, Kasilingam D, Costa AH. Speckle filtering of SAR images using Hölder regularity analysis of the sparse code. In: IEEE International Geoscience and Remote Sensing Symposium; 2003. pp. 63998-64000

[26] Tahiri AM, Farssi SM, Touzani A. Textures in Images Classification Using a Multifractal Approach. IEEE SITIS. 2005. Available from: http://www.u-bourgogne.fr/SITIS/05/download/Proceedings/Files/f138.pdf

[27] Bourissou A, Pham K, Levy-Vehel J. A multifractal approach for terrain characterization and classification on SAR images. IGARSS. 1994;**3**:1609-1611

[28] Jain AK, Murty MN, Flynn PJ. Data clustering: A review. ACM Computing Surveys. 1999;**31**(3):264-323

Chapter 7

Building an Integrated Database of Road Design Elements

Ali Dhafer Abed

Abstract

The road network is the main artery within the city structure, which requires designing of routes and classification within the standards. Hence, the importance of this chapter, which will focus on the standards and design elements of the engineering design of road in terms of road type system, functional classification system, traffic volume system, number of traffic lane system, road width design, side slopes and elevations of road layers, super elevation, design speed, overtaking and stopping sight distance, longitudinal and cross sections of the road path, design elements of horizontal and vertical curves, and intersections. The Civil 3D Land Desktop, GIS programs, and remote sensing technology will be used to design the path of major highway linking two urban areas in Mosul (Northern Iraq), which will be considered a case study. The path of the road and its elements will be designed according to special criteria that are compatible with the topography and nature of the area. The geometric data of the road will then be exported with all the design elements to the GIS program to build an integrated road database. The database is capable of spatial analysis and connectivity with other parts of the road network in the city.

Keywords: GIS, spatial database, road design, Civil 3D, AASHTO

1. Introduction

Transport has now become an important factor in determining the housing and workplaces of the largest segment of society. Transport has thus become an essential element in determining land uses; it is not affected by land uses only but affects them. The transport activities directly or indirectly lead to the transfer of civilization and civil landmarks to the farthest points in the country, open to human societies' paths to flow of science and knowledge, improve the health and social conditions and enjoy the joys of life and nature, and expand the perceptions of these communities and openness to ensure stability and development. In light of this, the efficiency of the transport system and roads requires continuous planning and design for the road, especially with respect to the design of engineering elements of the road, in order to provide the movement of society and maintenance at an acceptable level.

The engineering design of the road is defined as the process of finding the engineering dimensions of each road and arranging the visual elements of the road such as the path, distances of sight and passing, width, slopes, curves, super elevation, and other engineering characteristics. The horizontal and vertical design elements of the road are considered the most important elements of road design because their minimum limits are the basic for the design speed and the ruling slopes.

The horizontal and vertical roads' elements, the configuration of road path and grade affects safe operating speeds, and sight and passing distances for highway and roads' capacity establish the general character of a highway, more than any other design consideration. These components will significantly affect the safety, operational efficiency, and aesthetics of the highway.

Therefore, this chapter will focus on the design of the main elements of a road linking the city of Mosul and Makhmour (Northern Iraq) district south of Mosul. The surveying data was prepared by the Ministry of Municipalities for the proposed route of this road using a differential GPS device.

These surveys are a 900-m strip survey of the proposed route of the Mosul-Makhmour highway with a length of up to 20 km. Field surveying data will be used to prepare a geometric design for the proposed road linking for the two regions by using AutoCAD Civil 3D Land Desktop program and in accordance with international standards.

Initially, a path was proposed for this road depending on spatial analysis by GIS program to get optimum path linking between two areas. The stations, the type of road, the number of its lanes, its dimensions, its horizontal and vertical elements, and the slopes were defined according to the main design criteria of urban roads. The results were illustrated by longitudinal and cross-sectional plots showing the changes in the natural shape of the ground and all the elements of the horizontal and vertical curves of the road. After which, the volume of the earthworks was calculated for excavation and burial. Thus, a road connecting two urban areas was achieved in a manner that does not cause accidents and achieves the smooth flow of vehicles by making all elements of the road consistent with the expectations of drivers to avoid sudden changes in design specifications. Finally, all the designed road data were exported to the GIS program to build an integrated database for this road that can be linked to the rest of the city network, as well as all spatial analyses and network analyses.

2. Transportation planning

Transport is an important part of the planning process for cities and regions. Every planning activity, whether land uses, work centers, cultural, marketing, or leisure activities, depends in one way or another on transportation. The transport and traffic sector is considered an important sector in economic development, and this is reflected in the high expenses allocated to the development of this sector, which in Iraq up to 20% of the allocations of annual investment plans [1].

Transport planning is a structured approach to understanding traffic and transport characteristics. It aims to achieve an efficient and appropriate system that meets the current and future requirements of the community. We can define a number of objectives for this process [2]:

- Provide the most appropriate type of transport system according to the available potentials.
- Developing and increasing the efficiency of economic activity by reducing transport costs.
- Development of an integrated system of transport routes.
- Know how and when to improve the old road or build new roads according to future requirements.

- Optimal expenditures through the implementation of program cost and benefits for road projects to the general community.

- Development of programs and techniques for further urban and regional development.

- Reducing traffic accidents.

- To preserve and improve the environment.

- Also designed to design road networks according to planning standards, the most important of which are as follows [3]:

    - The network hierarchy, which is related to the functional classification of the different ways of fast, primary, and secondary traffic.

    - The network serves land use (residential, industrial, commercial, cultural, etc.) in a good way, providing easy access to parts of the city or regions.

    - The large and equivalent link between the city's internal network and the regional road network.

    - Network service for the gradual development of the city.

    - Avoid traffic jams that may occur at intersections by creating junctions at levels or any other planning solutions.

3. Road network classification systems

The systems adopted in the road network classification classify the network into four sections according to the following concepts [4]:

a. ***Road location systems and neighborhoods***: The road is classified according to its location according to the following concepts: the type of use of the road, the date of construction of the road, the uses of the ground surrounding the road, topography of the area of the road, and nature of the area.

b. ***Road engineering design system:*** These systems are based on the engineering classification of the road network according to the following systems, number of road lanes, system dividing the road, and switching systems from one road to another.

c. ***Road administration systems:*** These systems work on road management according to the following concepts, road planning according to levels, road paving systems and maintenance, road ownership systems, and type of tiling material systems.

d. ***Vehicle traffic systems:*** The road network is classified according to the movement of vehicles and their relation to the type of vehicle, according to the following, traffic volume of the road, purpose of the trip, type of vehicle using the road, and the roads' functional classification.

4. Classification of road network according to the criteria of capacity and rank

- Free streets: These streets are designed for the purpose of speeding and long distances in international and regional trips, with a capacity of between 1800 and 2000 vehicles/h, with lanes ranging from 4 to 8 lanes, at a speed of operation ranging from 80 to 120 km/h.

- Express streets: The streets are meant to serve the largest number of citizens, high speed, long distances for regional trips, a capacity of between 1400 and 1800 vehicles/h, and lanes of 4–8 lanes, with a speed of operation ranging from 60 to 80 km/h.

- Arterial streets: The streets with medium-distance urban trips, easy access between parts of the city and a capacity of 800–1200 km/h, with a speed of 40–60 km/h.

- Collected streets: These streets mean short urban trips, easy access between the city, and a capacity of 600–900 vehicles/h, with a speed of 30–50 km/h.

- Local streets: For short transport service, at a low speed of 20–30 km/h, with a capacity of 500–700 vehicles/h [5].

5. Functional classification of urban roads

The importance of functional classification is determined by which the role of each road is defined for the traffic and transport service. The degrees of urban roads vary according to the areas they serve, whether residential, commercial, residential-commercial, etc. and also according to the total movement that will be generated from those areas served. Classification of roads in urban planning can be summarized as follows [6–8]:

a. Major urban roads: These roads link the main centers of activity in urban areas. They are connected to the regional network and take the largest traffic load in the urban area. These roads have width about 40 m or more.

b. Secondary urban roads: These roads collect the vehicles from the main roads and distribute them to the degree of lower roads, and their widths are about (16–25 m).

Degrees	Minimum speed (km/h)	The desired speed (km/h)
Local road	30	50
Collector road	50	60
General arterial road	80	100
Less disturbance	70	90
Tangible disturbance	50	60
Highway	90	120

Table 1.
Design speed of urban roads [6, 7].

c. Urban roads of the third degree (local): Collecting vehicles from the residential areas and areas of activity to the highest road degree and carrying the least amount of traffic in the network and is the lowest degree in the hierarchy of the road network and its widths about (12–16 m).

The design characteristics of the road must be commensurate with the design speed chosen and expected for environmental and terrain conditions, and the designer should choose the appropriate design speed based on the planned road degree, terrain characteristics, traffic volume, and economic considerations. Note **Table 1**.

6. Highway capacity and level of service

The capacity of the road is the maximum number of vehicles expected to pass over a particular part of a lane or road during a given period of time under the prevailing traffic conditions.

Service level is the qualitative measurement of the effect of a number of factors such as operating speed, travel time, traffic failures and freedom of maneuver, and crossing, driving safety, comfort, road suitability, and operating costs for the service provided by the road to its users. **Table 2** shows the characteristics of the service level according to the type of road [9, 10].

The level of service	Urban arterial	Two lanes road
A	The average speed is about 90% of the speed of free flow. Delay at intersections with traffic signals is minimal	The average speed of the road is 93 km/h or greater. Most of the crossings are carried out without delay. In the ideal case, the traffic volume is 420 vehicles/h for both directions
B	The average speed of traffic decreases due to the delay in intersections and the impact of vehicles on some of them and about 70% of the speed of traffic. Load factor at 0.1 intersections and peak hour factor 0.8	The average speed of the road is 88 km/h or more. The load coefficient may be up to 0.27. Traffic volume is 750 vehicles/h for both directions
C	Travel speed is about 50% of the speed of free flow. Run balanced. Long rows of cars when traffic signals are possible	The average speed of the road is 84 km/h or more. The flow rate in the ideal case is about 43% of the capacitance, with a mean traffic in ideal conditions 1200 vehicles/h in both directions
D	Average speed 40% of free flow rate. The flow rate is unbalanced, and delays at intersections may be comprehensive	The average speed is 80 km/h. The flow rate is about 64% of the capacity, with continuity in the imposition of overflow and flow of approximately 1800 vehicles/h for both directions
E	Speed of flow is 33% of free flow speed, volume at capacity, and flow is not balanced. The coefficient of load at intersections (0.7–1.0). The peak hour factor is 0.95	The average flow rate is about 72 km/h. The flow rate in ideal conditions is 2800 vehicles/h; level E may not be accessible as the operation is converted from service level D to F directly
F	The average speed of traffic between 25 and 33% of the speed of free flow, high delay times at the branches of intersections	The operating speed is less than 72 km/h, and the traffic are overcrowded and constrained with unexpected characteristics, volume less than 2000 vehicles/h in both directions

Table 2.
Service level characteristics by road type [9, 10].

7. Specifications and determinants of roads' design and general criteria

7.1 Stopping sight distance (SSD)

The distance traveled by the vehicle from the instant the driver sights an object necessitating a stop to the instant the brakes are applied and the distance required to stop the vehicle from the instant brake application begin or defined as the sum of distances from when the driver decides to apply the break until the car stop, as in Eq. (1):

$$SSD = (0.278 \times V \times t) + (0.039 \times \frac{v^2}{a}) \tag{1}$$

where (SSD) is the stopping sight distance in m, (V) is the initial speed (kph), (a) is the rate of deceleration (3.4 m/S2), and (t) is the Brake reaction time, which is assumed to be 2.5 seconds by AASHTO [11].

Table 3 can illustrate the stopping sight distance and its relation to the design speed, brake reaction distance, and braking distance on level. **Table 4** illustrates increment of the stopping sight distance and its relation to the design speed and brake in state of slope directed down [12–14].

Design speed (Km/h)	Brake reaction distance (m)	Braking distance on level (m)	Stopping sight distance	
			Calculated (m)	Design (m)
20	13.9	4.6	18.5	20
30	20.9	10.3	31.2	35
40	27.8	18.4	46.2	50
50	34.8	28.7	63.5	65
60	41.7	41.3	83.0	85
70	48.7	56.2	73.4	129
80	55.6	73.4	129.0	130
90	62.6	92.9	156.5	160
100	59.5	114.7	184.2	185
110	76.5	138.8	215.3	220
120	83.4	156.2	248.5	250
130	90.4	193.8	284.2	285

Table 3.
Stopping sight distance and its relation to the design speed and brake [12].

Design speed (km\h)	Increase the stopping sight distance in state of downslope (m)		
	3%	6%	9%
40	2	4	6
50	3	6	10
60	5	10	18
70	7	15	6
80	9	21	-
90	12	29	-
100	16	38	-

Table 4.
Relationship between design speed and increment of stopping sight distance in state of downslope [15].

Design speed (km/h)	Assumed speed (km/h)		Passing sight distance (m)	
	Passed vehicle	Passing vehicle	From exhibit 3–6	Rounded for design
30	29	44	200	200
40	36	51	266	270
50	44	59	341	345
60	51	66	407	410
70	59	74	482	485
80	65	80	538	540
90	73	88	613	615
100	79	94	670	670
110	85	100	727	730
120	90	105	774	775
130	94	109	812	815

Table 5.
Passing sight distance for the design of two-lane highways [16, 17].

7.2 Passing sight distance (PSD)

It represents enough free distance of traffic so that the driver can see the driver in front of him to be able to complete the process of circumventing without touching the car that passes without being intercepted by any counter vehicle may appear after the start of the bypass and then return to the right warm easily after the overtaking process. PSD is designed for two-lane highway as in **Table 5** which illustrate passing sight distance with respect to the design speed for passed and passing vehicle [16].

7.3 Horizontal planning of the road

The horizontal curve is a part of circular curves, which consist of intersection of two tangents of road. Horizontal curve has four types illustrated in **Figure 1**. The location and configuration of the horizontal curve are affected by some of the factors as follow [18]:

1. Physical condition: land uses, earth topography and geophysical conditions, intersection with waterway and man-made barriers.

2. Environmental circumstances: impacts on the adjacent land use, community-based impacts, and environmentally sensitive areas.

3. Economics condition: cost of construction, road ownership costs, effects of utility, costs of operating, and maintenance.

4. Road safety: distance of sight, alignment consistency, considerations of the human factor.

5. Classification and design considerations of highway: level of service, functional classification, design speed, and standards.

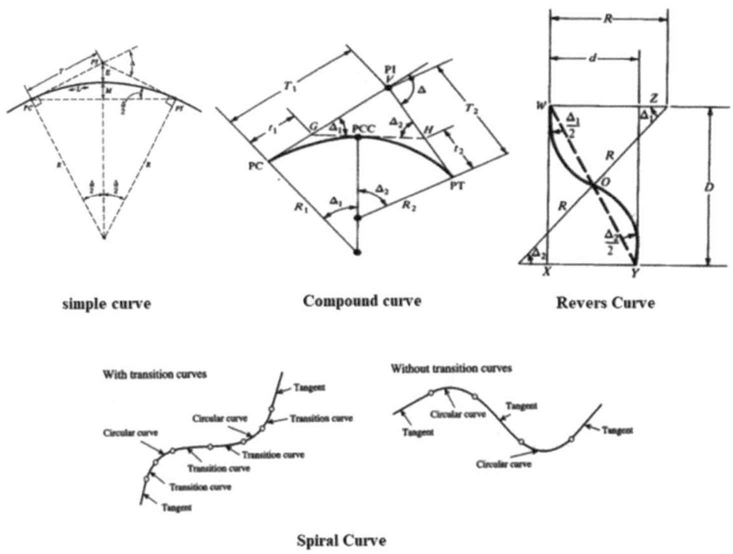

Figure 1.
Types of horizontal curves.

7.4 Vertical planning of the road

Vertical road planning consists of a series of longitudinal tendencies connected to each other by vertical curves (note **Figure 2**). Vertical planning is governed by a number of factors: safety, terrain, road speed, design speed, horizontal planning, construction cost, vehicle characteristics, and rain drainage. Visibility in all parts of the longitudinal sector must be met with the minimum distance required to stop (not overtaking), according to the design speed corresponding to the roadway. There are general considerations in the vertical planning of the road, which can be summarized as follows [19, 20]:

1. The goal should be to obtain an easy linear elevation design with gradual changes in line with the type of road or its degree and the nature of the land.

2. Avoid wavy vertical planning or vertical planning with hidden dips, because it is bad-looking and dangerous. Hidden dips cause accidents in overtaking, fooling the overtaking driver beyond the low, and thinking the road is free of anti-cars. But in the low-depth depressions, such as a longitudinal ripple, there is a lack of reassurance in the driver because it cannot determine the presence or absence of a vehicle likely to be hidden behind the high part. This type of longitudinal layout can be avoided by horizontal curvature or gradual change of slopes at light rates, possibly by increasing cutting and filling.

3. The longitudinal refraction bending planning should be avoided (two vertical curves in the same direction separated by a short tangent), especially in concave curves where the complete view of the two curves is not acceptable.

4. It is preferable for long slopes to have steep slopes at the bottom, and then the slope falls close to the top, or the continuous gradient is reduced by the introducing short distances where the slope is less than that of a regular full slope. This is especially relevant for low-design speed road conditions.

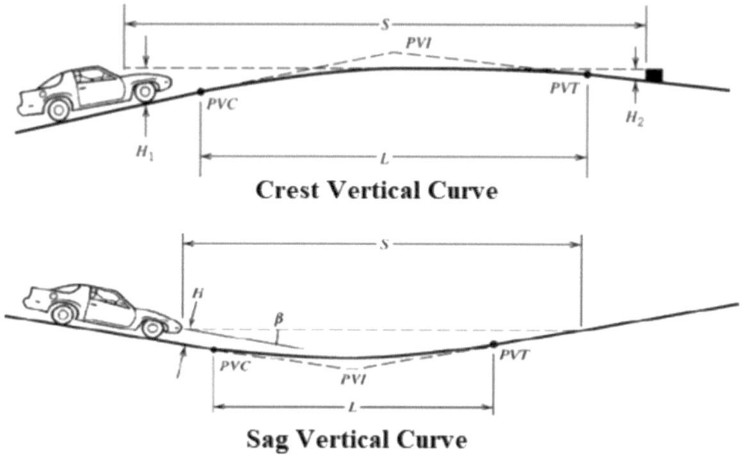

Figure 2.
Type of vertical curve [19].

5. K values can be used to compute the length of vertical curve for the crest and sag vertical curves. And vertical curve should have minimum length equal to three times the design speed.

6. SSD in most cases will be used for the length of vertical design, but for trucks it is not necessary because the driver of the truck is able to see farther than the passenger car. So, the SSD for trucks and passenger cars is balance.

Table 6 can illustrate the relationship between design control for SSD with respect to the K value for the vertical curve, while **Table 7** shows design controls for vertical curve based on PSD [21].

Design speed (km/h)	Stopping sight distance (m)	Rate of vertical curvature (K)	
		Calculated	Design
20	20	0.6	1
30	35	1.9	2
40	50	3.8	4
50	65	6.4	7
60	85	11	11
70	105	16.8	17
80	130	25.7	26
90	160	38.9	39
100	185	52	52
110	220	73.6	74
120	250	95	95
130	285	123.4	124

Table 6.
Design control for stopping sight distance with respect to the K value for vertical curve [20].

Design speed (km/h)	Passing sight distance (m)	Rate of vertical curvature (K) design
20	200	46
30	270	84
40	345	138
50	410	195
60	485	272
70	540	338
80	615	438
90	670	520
100	730	617
110	775	695
120	775	695
130	815	769

Table 7.
Design controls for vertical curve based on passing sight distance [21].

7.5 Super elevation

Super elevation allows the car to travel across a curve safely and at a higher speed than is possible with the natural crown section. The overall super elevation rate increases with speed and a sharper curvature (note **Figure 3**). **Table 8** can illustrate the maximum lateral lifting value of super elevation [22].

where Rv is the vehicle's traveled path radius, Ff is the force of side frictional, FC is the centripetal force, Wp is the weight of vehicle parallel to the road path surface, W is the vehicle weight, Wn is the vehicle weight normal to the road path surface, Fcn is the gravitational force that works naturally on the road surface, e is the number of vertical of rise per one horizontal station (100 m), and α is the incline angle [23].

7.6 Side slope of cut and fill

Side slopes are designed to ensure the stability of the road and to provide the opportunity to secure cars out of control. **Table 9** shows the relationship between

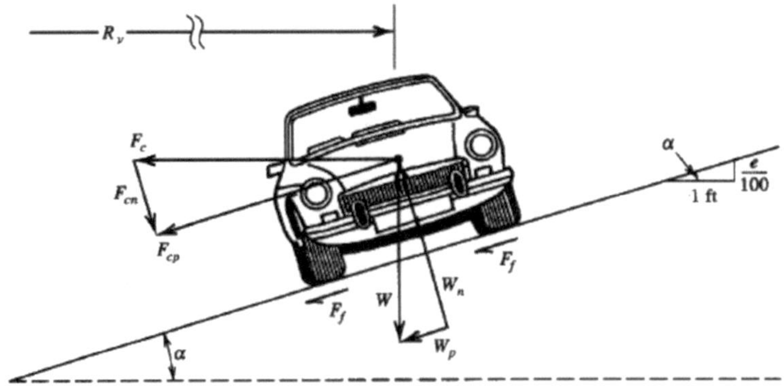

Figure 3.
Super elevation [22].

Degree of the road	Maximum side lifting value of the road is desirable (m/m)	Maximum lateral lifting value is absolute (m/m)
Highway	0.08	0.10
Arterial road	0.08	0.10
Collector road	0.08	0.12
Local road	0.10	0.12

Table 8.
Maximum lateral lifting value according to AASHTO [22].

Height (m)	Earth work	Plan		Wavy		Mountainous	
		Desired	Max slop	Desired	Max slop	Desired	Max slop
0–1	Cut	1:6	1:4	1:6	1:3	1:6	1:3
	Fill	1:6	1:6	1:4	1:4	1:4	1:4
1–3	Cut	1:4	1:3	1:3	1:2	1:3	1:2
	Fill	1:4	1:4	1:4	1:4	1:3	1:3
5–3	Cut	1:3	1:2	1:3	1:2	1:3	1:2
	Fill	1:4	1:3	1:4	1:3	1:3	1:1.5
5	Cut	1:2	1:2	1:2	1:2	1:2	1:2
	Fill	1:3	1:2	1:3	1:2	1:2	1:1.5

Table 9.
Side slope (horizontal to vertical) for the type of terrain except rocks [25].

the topography type and the height of the cutting or the filling, and the maximum side slope desired in the roads for the filling slope less than or equal to (2:1) depends on soil analysis [24, 25].

8. Spatial data of study area and method of processing

For the purpose of designing the proposed road elements of the study area between Mosul and Makhmour, the spatial data of the study area were obtained from the Ministry of Municipalities of Mosul City.

The spatial data is the field survey data of the route of the road, completed by a team of engineers from the Ministry of Municipalities. The survey data was conducted in the form of a strip width 900 m around the proposed route. A 900-m width was selected to cover all the places that the road path might pass, because the path was selected roughly, not accurately. The length of the strip survey is 20 km to connect the two urban areas.

The survey data contains a set of point coordinates (3626 points) observed with a high-precision equipment (Leica viva GS15). The coordinates' projection was WGS84-UTM-Zone38N. These data is an unprocessed raw data, unrelated to each other, and contains many coordinates that may not be connected to the pathway. For the purpose of processing these data and linking it together, adjusting the system of coordinates, adjusting the elevations, and creating a digital elevation model for the region, the Civil 3D Land Desktop program will be used to process this data and then export it to the GIS program.

Spatial analysis will be used in the GIS program to select the optimal path that connects the two study areas based on spatial data. The optimal path for the proposed road will be chosen according to planning and design criteria. This path will be exported to the Civil 3D program again to identify the rest of the design elements of the road.

In the Civil 3D program, all the design elements of the road will be defined according to the AASHTO standards, leading to the final stage where the road contains an integrated database ready for implementation.

9. Methodological steps for designing road elements using Civil 3D Land Desktop and GIS software

In order to define the proposed road path and all its horizontal and vertical and design elements, these points will be used in the definition and design of road elements according to the following methodological steps:

a. The civil 3D land desktop program has been configured to be in meter units, 1:1000 horizontal scale and 1:50 vertical scale, UTM, WGS 84 Datum as **Figure 4**.

b. The survey points were imported within the program and arranged and modified (point number, coordinates values, elevations) to be ready for the purpose of design road elements, as in **Figure 5**. Constructing a surface triangulated irregular network (TIN) and connecting the points' coordinates for the purpose of deriving the elevations of the road depending on it, as in **Figure 6**. The TIN surface is then exported to GIS as digital elevation model (DEM).

c. The Geographic Information system (GIS) program was used to select the best route for the road connecting the city of Makhmour with the city of Mosul, as in **Figure 7** which shows the sequence of steps to choose the path. The network of TIN has been reclassified according to the elevation values, so that the road passes from the flat areas as much as possible, as the land on which the road passes has steep slopes. The land use layer has been reclassified so as not to cross the road with any unwanted land use. The best and shortest route between the two urban areas was then chosen. The length of the best selected path was 18210.88 m. Its alignment, stations, width, and number of lanes were defined and exported to the civil 3D.

d. ASHTO standards were adopted in the design of horizontal elements of the proposed highway. These design elements were defined to the Civil 3D as follows:

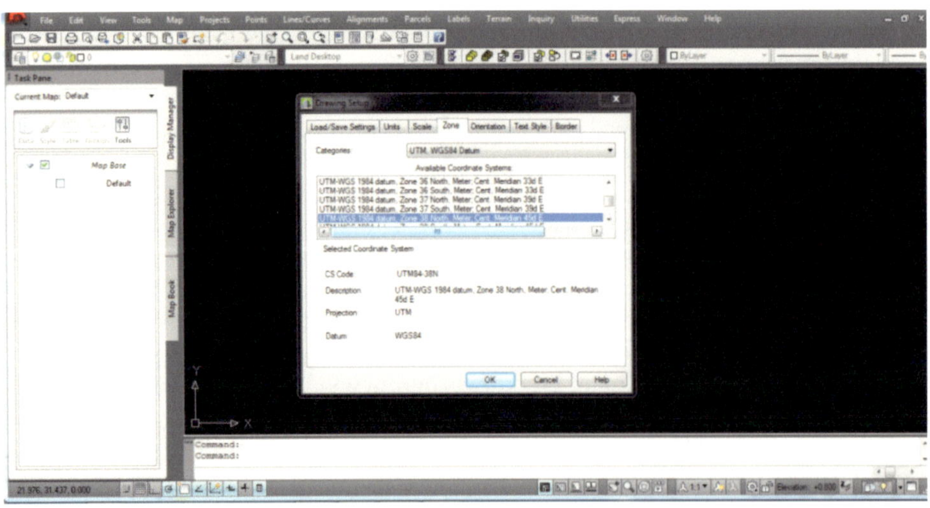

Figure 4.
Configuration of units, scale, projection of Civil 3D program.

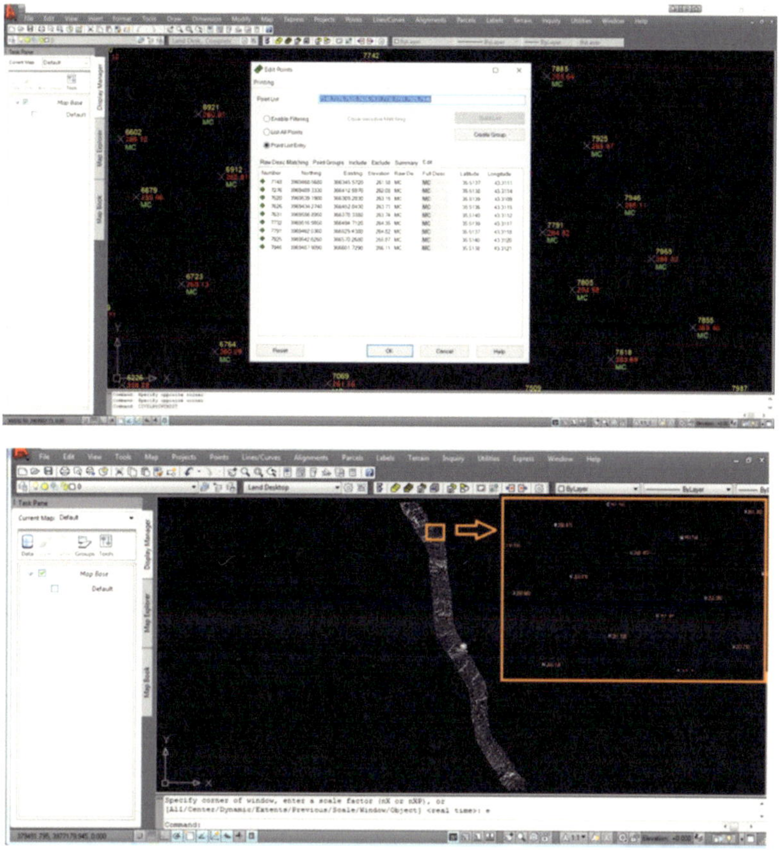

Figure 5.
Modify coordinates of survey points.

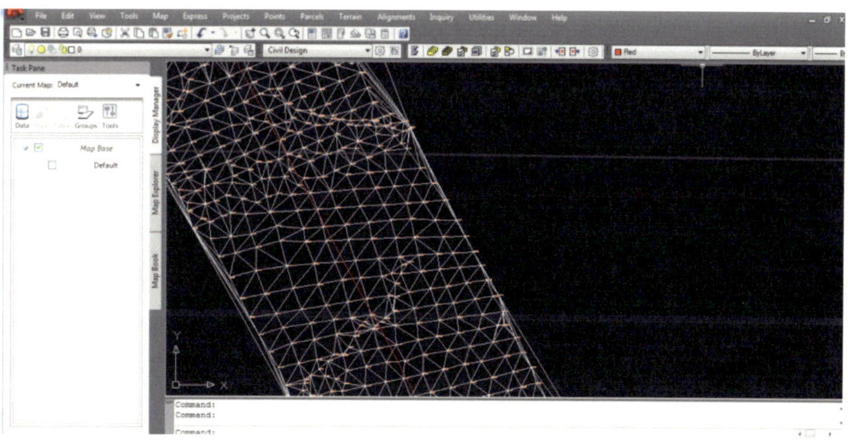

Figure 6.
Constructing the TIN surface to link between coordinates.

- The design speed = 100 km/h.
- Maximum longitudinal slope = 3%.
- Maximum side slope = 4/1.
- Stopping sight distance SSD = 185 m.

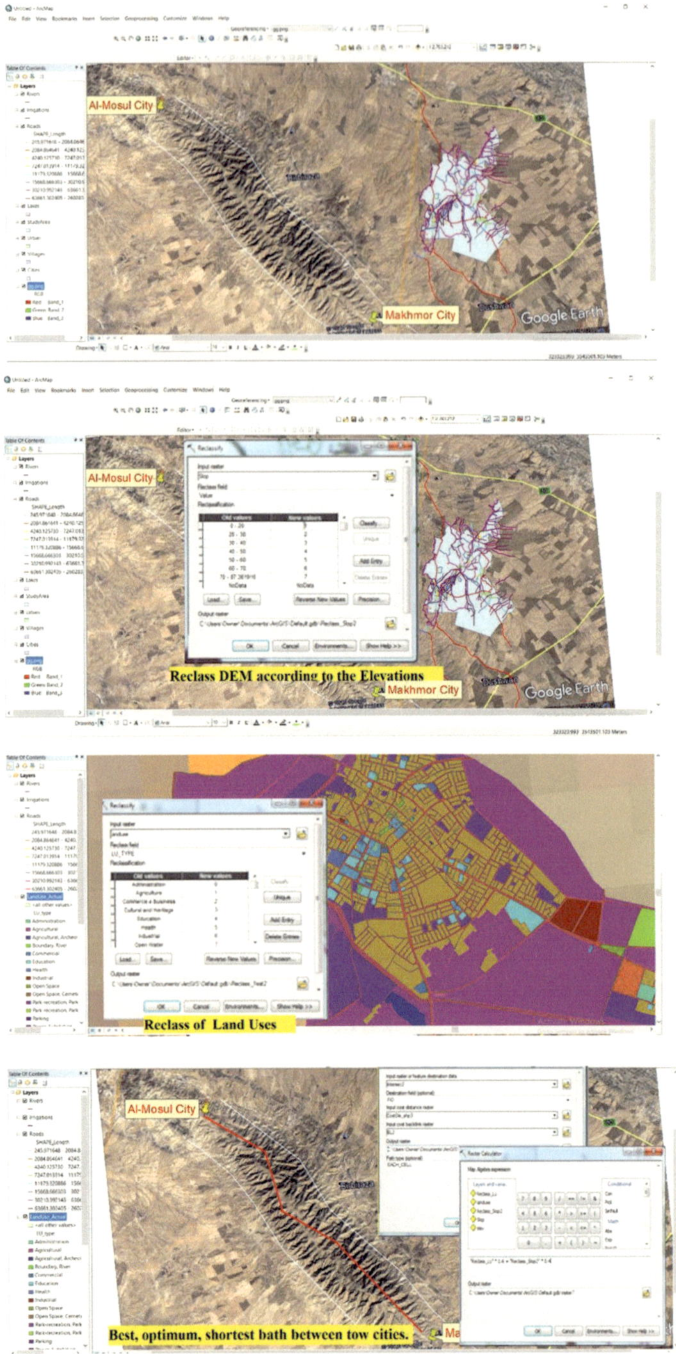

Figure 7.
Sequence of steps in GIS to select the best route for connecting the city of Makhmour with the city of Mosul.

- Supper elevation = 4%.
- Break reaction distance = 60 m.
- Break distance on level = 115 m.
- For concave and convex vertical curves (sag and crest), SSD = 185 m, PSD = 670 m, K = 45.

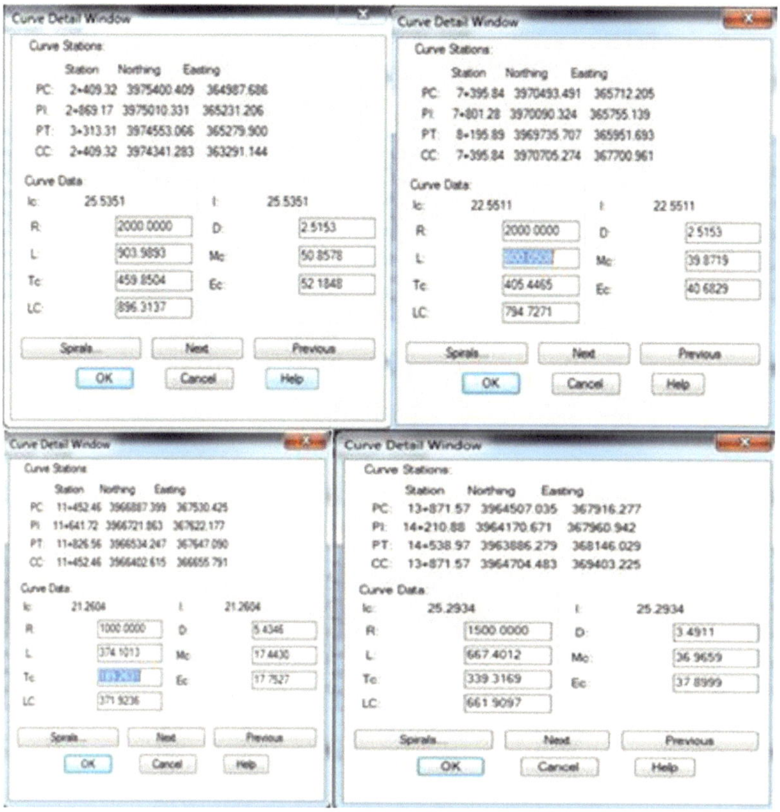

Figure 8.
Design elements for horizontal curves.

- Minimum length of vertical curve = 700 m.
- The level of service was A, the number of lane 2 was with two direction, the chosen width was 12 m per lane, and the traffic volume was 420 vehicles/h in both directions.

e. The design elements for the horizontal curves were illustrated as in **Figures 8** and **9**. Where the highway contained four horizontal curves designed according to the design standards. **Figure 10** can illustrate the proposed path of road after adding the stations. While, **Figure 11** illustrate the definition of the design velocity of horizontal curves.

f. A longitudinal section of the proposed road was produced on a horizontal scale (1:600) and a vertical scale (1:100), as in **Figure 12a** and **b**.

g. The construction line (formation level), which represents the final level of the road, has been defined so that it achieves the lowest proportion of cuttings and fill in the earthworks and with the lowest vertical curves and as in **Figure 13**, which illustrates the elements of vertical curves and stations of PVC, PVI, and PVT through the table stations, as well as the levels of these stations, their slopes, and the length of their vertical curves, and the type of curve. **Figures 14** and **15** illustrate the longitudinal section after adding the line of construction and vertical curves.

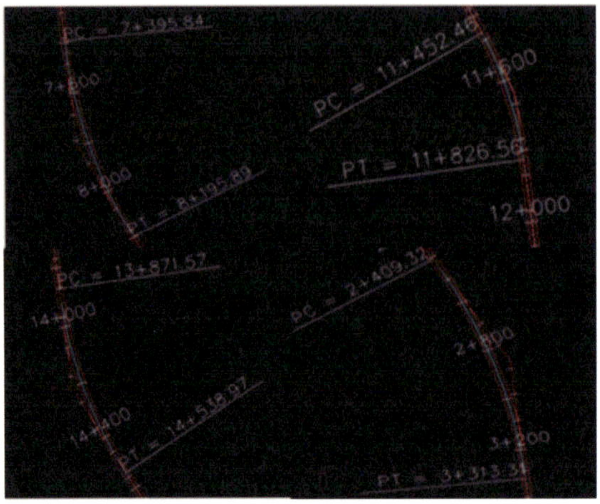

Figure 9.
Design elements for the horizontal curves.

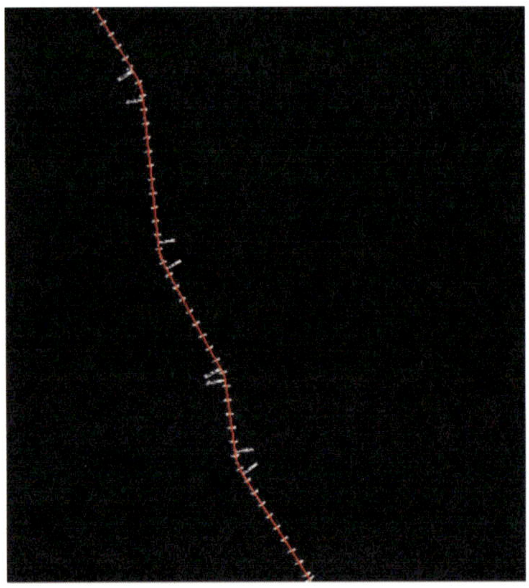

Figure 10.
The proposed path of highway after adding the stations.

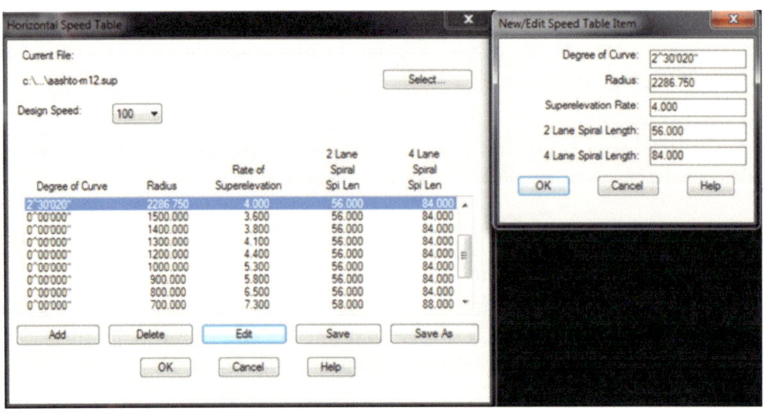

Figure 11.
Define design velocity of horizontal curves.

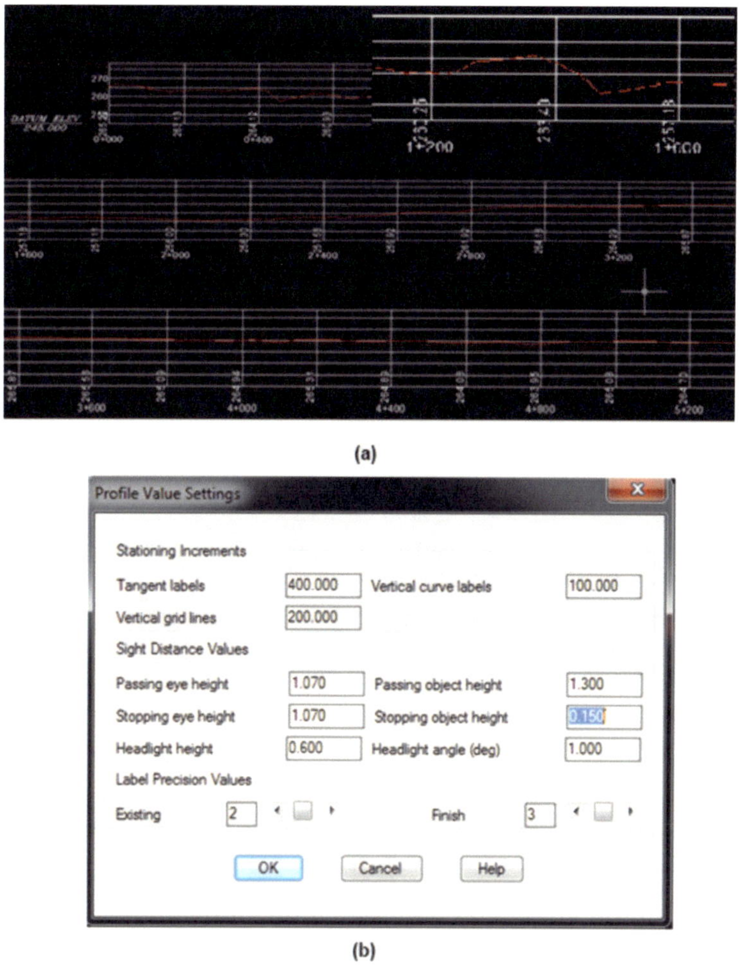

Figure 12.
(a) Longitudinal section of the proposed road and (b) profile's design elements.

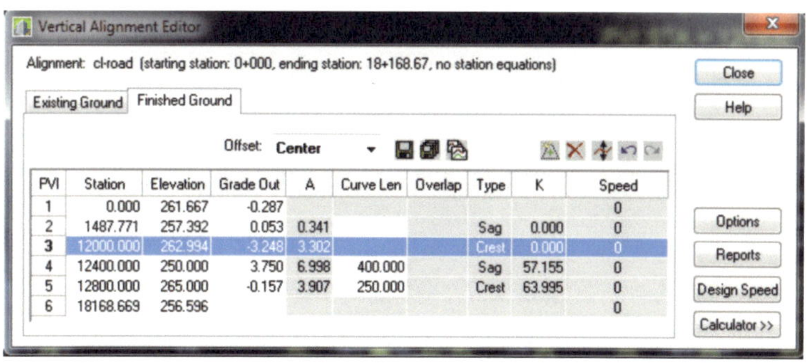

Figure 13.
Define vertical curve elements of the road.

h. The cross sections of the proposed road were generated and designed according to the criteria that correspond to the cross sections of the reality of the case (width of the total road 24 m, shoulder length 1 m, side slope 1/2, supper elevation 4%, the thickness of the tiling was defined as 10 cm, thickness of the sub-base 15 cm). Note **Figure 16** which illustrates the forms of some of cross sections.

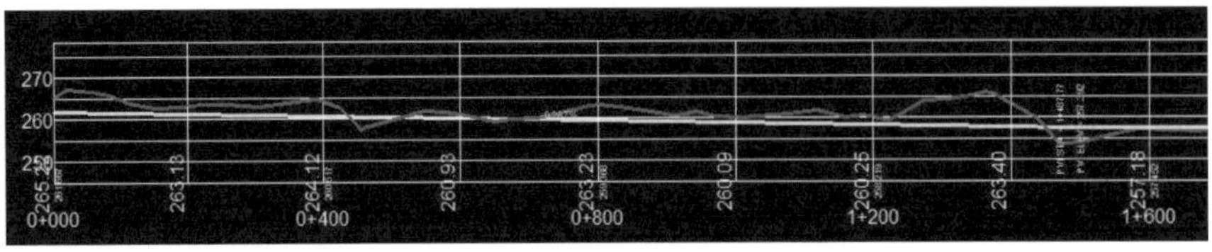

Figure 14.
The longitudinal section of the road after adding construction line.

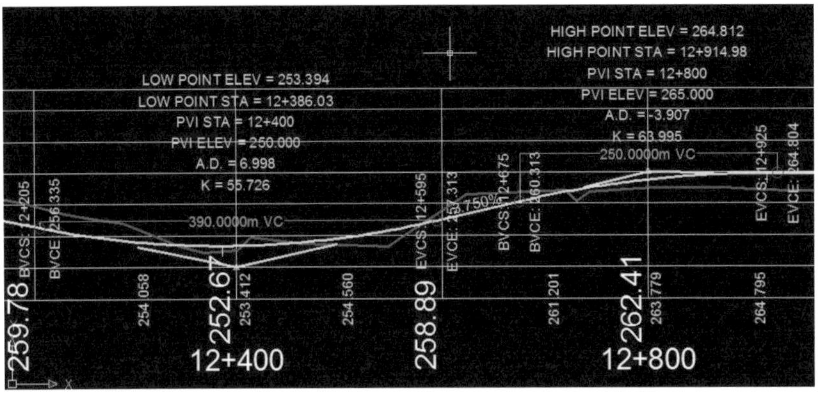

Figure 15.
Section of profile illustrates vertical curve elements.

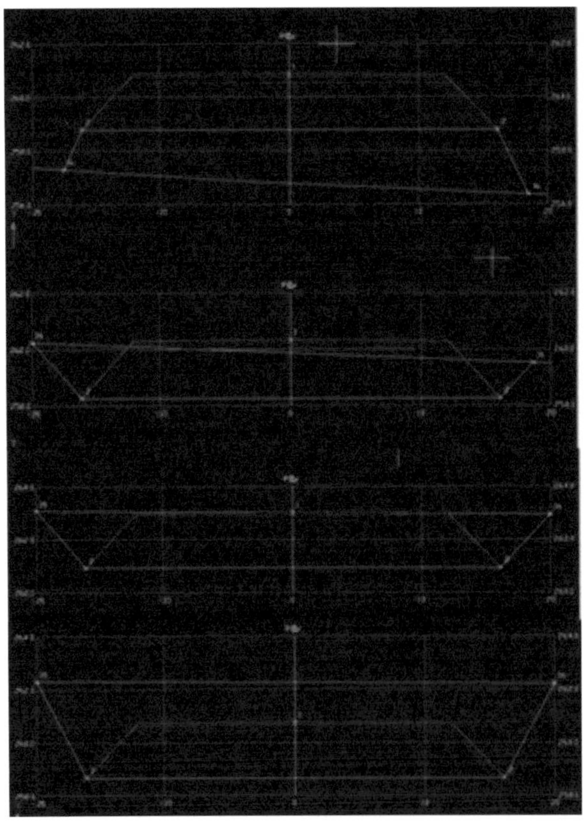

Figure 16.
Forms of some cross sections of the proposed road.

i. Calculated volume of the earth works in a way of prismoidal formula as shown in **Table 10**, which shows the volume of cut and fill for each station of the proposed road. These data of volume were exported from Civil 3D to GIS as attribute table.

STATION	AREAS Square Meters		VOLUMES Cubic Meters		CUMULATIVE VOLUMES Cubic Meters	
	CUT	FILL	CUT	FILL	CUT	FILL
0+000	145.151	0.000				
0+200	90.603	0.000	23575.485	0.000	23575.485	0.000
0+400	147.837	0.000	23844.042	0.000	47419.527	0.000
0+600	52.267	0.000	20010.439	0.000	67429.965	0.000
0+800	156.472	0.000	20873.955	0.000	88303.921	0.000
1+000	63.893	0.000	22036.510	0.000	110340.431	0.000
1+200	90.299	0.000	15419.150	0.000	125759.581	0.000
1+400	224.767	0.000	31506.509	0.000	157266.091	0.000
1+600	13.054	6.880	23782.063	688.023	181048.154	688.023
1+800	10.724	10.904	2377.797	1778.454	183425.950	2466.478
2+000	1.215	45.127	1193.871	5603.179	184619.821	8069.657
2+200	1.474	39.404	268.920	8453.158	184888.741	16522.815
2+400	12.326	7.576	1380.001	4698.038	186268.742	21220.853
2+600	86.949	0.000	9943.715	756.838	196212.457	21977.691
2+800	156.671	0.000	24367.669	0.000	220580.127	21977.691
3+000	232.884	0.000	38963.356	0.000	259543.483	21977.691
3+200	233.013	0.000	46597.423	0.000	306140.906	21977.691
3+400	286.345	0.000	51943.413	0.000	358084.319	21977.691
3+600	271.226	0.000	55757.114	0.000	413841.433	21977.691
3+800	250.393	0.000	52161.953	0.000	466003.386	21977.691
4+000	241.469	0.000	49186.241	0.000	515189.627	21977.691
4+200	250.824	0.000	49229.366	0.000	564418.992	21977.691
4+200	250.824	0.000	49229.366	0.000	564418.992	21977.691
4+400	231.797	0.000	48262.131	0.000	612681.124	21977.691
4+600	199.397	0.000	43119.390	0.000	655800.513	21977.691
4+800	190.153	0.000	38954.983	0.000	694755.497	21977.691
5+000	227.025	0.000	41717.828	0.000	736473.325	21977.691
5+200	209.558	0.000	43658.352	0.000	780131.677	21977.691
5+400	246.274	0.000	45583.177	0.000	825714.853	21977.691
5+600	224.133	0.000	47040.673	0.000	872755.527	21977.691
5+800	213.506	0.000	43763.894	0.000	916519.420	21977.691
6+000	183.222	0.000	39672.745	0.000	956192.165	21977.691
6+200	190.872	0.000	37409.391	0.000	993601.556	21977.691
6+400	174.304	0.000	36517.660	0.000	1030119.215	21977.691
6+600	48.769	0.000	22307.314	0.000	1052426.528	21977.691
6+800	181.369	0.000	23013.761	0.000	1075440.290	21977.691
7+000	10.700	11.373	19206.862	1137.284	1094647.152	23114.976
7+200	75.335	0.000	8603.477	1137.284	1103250.629	24252.260
7+400	70.033	0.000	14536.798	0.000	1117787.427	24252.260
7+600	72.421	0.000	14240.503	0.000	1132027.929	24252.260
7+800	71.390	0.000	14378.283	0.000	1146406.213	24252.260
8+000	0.000	126.003	7138.767	12602.797	1153544.980	36855.056
8+200	0.133	68.289	13.264	19437.703	1153558.244	56292.759
8+400	0.012	63.891	14.526	13217.996	1153572.770	69510.756
8+600	0.338	53.595	35.025	11748.638	1153607.795	81259.393
8+800	5.087	24.576	542.486	7817.149	1154150.280	89076.543
9+000	2.756	33.450	784.315	5802.622	1154934.596	94879.165
9+200	0.001	64.138	275.759	9758.794	1155210.355	104637.959
9+400	9.443	13.495	944.438	7763.315	1156154.793	112401.273
9+600	3.939	28.133	1338.242	4162.873	1157493.035	116564.146
9+800	4.911	24.861	884.978	5299.469	1158378.013	121863.614
10+000	5.262	24.130	1017.225	4899.184	1159395.239	126762.798
10+200	0.000	112.118	526.163	13624.804	1159921.402	140387.602
10+400	0.000	133.695	0.000	24581.235	1159921.402	164968.837
10+600	0.000	90.933	0.000	22462.803	1159921.402	187431.641
10+800	0.000	68.839	0.000	15977.188	1159921.402	203408.829
11+000	0.000	103.706	0.000	17254.468	1159921.402	220663.296
11+200	0.000	157.456	0.000	26116.178	1159921.402	246779.474
11+400	0.000	141.648	0.000	29910.354	1159921.402	276689.827
11+600	0.000	171.626	0.000	31316.229	1159921.402	308006.056
11+800	4.219	28.558	423.992	20001.096	1160345.394	328007.152
12+000	17.092	0.091	2131.032	2864.867	1162476.425	330872.019
12+200	135.411	0.000	15250.317	9.069	1177726.743	330881.088
12+400	14.049	13.930	14946.010	1393.011	1192672.753	332274.100
12+600	66.798	0.000	8084.698	1393.011	1200757.451	333667.111
12+800	2.149	36.350	6894.742	3635.028	1207652.193	337302.139
13+000	0.000	79.201	214.906	11555.121	1207867.099	348857.260
13+200	0.641	45.393	64.076	12459.372	1207931.175	361316.632
13+400	0.752	44.625	139.251	9001.733	1208070.427	370318.365
13+600	0.094	54.466	84.598	9909.067	1208155.024	380227.432
13+800	2.238	35.127	233.209	8959.320	1208388.233	389186.751
14+000	0.000	66.148	223.562	10128.635	1208611.795	399315.386
14+200	0.000	59.495	0.000	12564.774	1208611.795	411880.160
14+400	0.000	63.980	0.000	12349.121	1208611.795	424229.281
			0.000	13171.003	1208611.795	437400.374

Table 10.
Calculated volume of cut and fill for each station of the proposed road.

10. Conclusion

1. The length of the proposed road was 18210.88 m according to the data of the Ministry of Municipalities, 24 m width with two corridors, and the coordinates system was UTM.WGS 1984 and Zone 38N.

2. The TIN is the basis of the accuracy of the roads' coordinates, because it is observed with accurate GPS devices.

3. All horizontal and vertical road elements are defined through the CIVIL 3D program, facilitating and accelerating the design process in accordance with international standards.

4. Four horizontal curves and three vertical curves were proposed for the proposed road, according to the topography of the earth, which required this number of curves.

5. The design of the roads using Civil 3D and GIS in all its components makes the design process highly efficient through the speed of time, little effort, and low cost.

6. Civil 3D has all the international standards used in road design and has all the tools that can easily define all design elements for roads and export it to GIS.

7. The program provides us with longitudinal and cross sections that show the change in ground and construction line levels very accurately, which facilitates the process of proposing tiling and cladding levels.

8. The volume calculated by using the program is very precise. The earthwork produced by the pieces can be used to bury the areas that need to be buried. The construction line chosen to represent the proposed road level was chosen at the same depth as the depth of the burial.

9. The road data exported to the GIS program has created an integrated road database. This database can be performed on any kind of spatial analysis or network analysis of the roads within the environment of the GIS.

Acknowledgements

The authors would like to thank Mustansiriyah University (www.uomustansiriyah.edu.iq) Baghdad, Iraq, for its support in the present work.

Author details

Ali Dhafer Abed
Civil Engineering Department, College of Engineering, Mustansiryah University, Baghdad, Iraq

*Address all correspondence to: ali.dhafer@uomustansiriyah.edu.iq

References

[1] Ahmed MF. Hierarchical hierarchy of roads and spatial organization [master thesis]. Urban and Regional Planning Center. Baghdad University; 2010. p. 52

[2] Al-Hamdani, Mehdi W. Building a future model for traffic volumes for sectors in Ramadi [master thesis]. Introduction to Faculty of Engineering, Anbar University; 2016. p. 112

[3] The narrator, Falih MK. The impact of the development of road network in the urban growth of the city of Rawa [Master thesis]. Urban and Regional Planning Center, Baghdad University; 2002. pp. 33-34

[4] Abed AD. Planning and design of highways according to AASHTO standards using remote sensing technology (Samarra City as a case study). In: ICOASE. IEEE. 2018. pp. 3-4. DOI: 10.1109/ICOASE.2018.8548823

[5] Al-Fahdawi MS. Performance of the transport system and models of land use for the city of Ramadi [a master thesis that is not published]. Introduction to Faculty of Engineering, Anbar University; 2003. p. 200

[6] Hassoun Morjan DR. Planning support for decision makers using GIS [unpublished PhD thesis]. Submitted to the Higher Institute of Urban and Regional Planning; 2015. pp. 11-13

[7] General Authority for Roads and Bridges, Iraq. The classification of the road network in Iraq at the country level. Unpublished data. 2019. pp. 55-60.

[8] Tolley R, Brian T. Transportation System, A Geographical Approach. London: Prentice, Longman; 1999

[9] www.proposal for the classification of road network facilities and assignment of delay functions in GTA EMME/2 (General Technical Application networks)

[10] Post-Positivism and Beyond: GIS and Urban Studies. Urban Geography. No. 15. 2018

[11] Highway Department. Guidance notes road inspection manual. https://www.hyd.gov.hk/en/publications_and_publicity/publications/technical_document/guidance_notes/pdf/GN016C.pdf

[12] Lorenso I. Spatial data analysis and Gis applied to study the urban

[13] Darton R. Rotation in Factor Analysis Statistication. 1989. pp. 263-375

[14] Harman. Modern factor analysis. USA

[15] Geographic information science implications for urban and regional planning. URISA Journal. 2012. https://pdfs.semanticscholar.org/5d15/374050798f0c4274444bb2b793453d4a5919.pdf

[16] www.Ema S. GIS applications for capturing managing and publishing Geomediabases. 2000

[17] www.Andrew G. Macbeth. Road Classification Systems. Christchurch and Toronto. September 2001

[18] www.Ana R. Application of GIS in Land-use. Anselin, Luc: GIS Research Infrastructure for Spatial Analysis of Real Estate Markets; 2015

[19] Anselin L. Geographic information systems (GIS) in housing and mortgage finance. Journal of Housing Research. 1998;**9**(1):113-133

[20] A Policy on Geometrical Design of Highways and Streets. AASHTO; 2001. http://www.math.wichita.edu/~axmann/Math714/TrafficCircles/Bosworth1i

[21] Design Standards, Ministry of Communications: Government of the People's, Republic of Bangladesh; 2012

[22] Highway Agency. Design Manual for Roads and Bridges. Vol. 6. UK: Road Geometry; 2017

[23] American Association of State Highway and Transportation Officials (AASHTO). Washington, DC; 2001

[24] Road and Highway Department. Road Materials and Standards. Final Report. Vol. 11. Development of Geometric. Bangladesh; 1999

[25] Ogden KW. Safer Roads: A Guide to Road Safety Engineering. UK Library. 1st ed. UK; 1996

Chapter 8

On the Use of Low-Cost RGB-D Sensors for Autonomous Pothole Detection with Spatial Fuzzy *c*-Means Segmentation

Yashon Ombado Ouma

Abstract

The automated detection of pavement distress from remote sensing imagery is a promising but challenging task due to the complex structure of pavement surfaces, in addition to the intensity of non-uniformity, and the presence of artifacts and noise. Even though imaging and sensing systems such as high-resolution RGB cameras, stereovision imaging, LiDAR and terrestrial laser scanning can now be combined to collect pavement condition data, the data obtained by these sensors are expensive and require specially equipped vehicles and processing. This hinders the utilization of the potential efficiency and effectiveness of such sensor systems. This chapter presents the potentials of the use of the Kinect v2.0 RGB-D sensor, as a low-cost approach for the efficient and accurate pothole detection on asphalt pavements. By using spatial fuzzy *c*-means (SFCM) clustering, so as to incorporate the pothole neighborhood spatial information into the membership function for clustering, the RGB data are segmented into pothole and non-pothole objects. The results demonstrate the advantage of complementary processing of low-cost multisensor data, through channeling data streams and linking data processing according to the merits of the individual sensors, for autonomous cost-effective assessment of road-surface conditions using remote sensing technology.

Keywords: Kinect RGB-D sensor, pothole detection, spatial fuzzy-*c* means clustering (SFCM), sensor calibration

1. Introduction

Presently, two approaches are typically used to monitor the condition of pavements: manual distress surveys and automated condition surveys using specially equipped vehicles. Traditionally, in order to determine the serviceability of road pavements, designated pavement officers perform on-site inspection, either by walk-observe-record or by windshield (drive-by) inspection, so as to aggregate the roughness, rutting and surface distresses [1, 2]. With the advancement of sensor technology, numerous automatic pavement evaluation systems have been proposed to aid in pavement condition inspection during the last two decades [3]. Currently, there exist several off-the-shelf commercial systems, which are being widely used by some of the road maintenance agencies for detailed pavement distress evaluation and exclusive crack analysis. Among which, the Fugro Roadware's ARAN, CSIRO's RoadCrack and Ramböll OPQ's PAVUE are of the world's leading manufacturers offering an integrated full-fledged pavement evaluation system equipped with Global Positioning System (GPS)/Inertial Measurement Unit (IMU) sensors, Light Detection And Ranging (LiDAR) system, high definition video camera, and special lighting illumination systems [2]. Nonetheless, technology for the monitoring of pavement condition does not appear to have kept pace with other technological improvements over the past several years. Furthermore, these pavement monitoring and evaluation approaches remain rather reactive than proactive in terms of detecting distresses and damage, since they merely record the distress that has already appeared, and most of these methods either require significant personnel time or the use of costly equipment. Thus these systems and techniques can only be used cost-effectively on a periodic and or localized basis, and may not allow for continuous long-term monitoring and deployment at the network level, due limitations in hardware and software development and costs.

For sustainable and cost-effective road infrastructure management, the road agencies charged with the responsibility of road maintenance and repairs should be able to continuously collect road condition data within their network, with the objective of building and implementing pavement information and management systems (PIMS) using non-destructive techniques. However, as already stated above, data collection for a whole network such as an entire city or town is expensive and time consuming, if pursued by traditional surveys. Developments in sensor technology for digital image acquisition and computer technology for image data storage and processing can allow the local agencies to use digital image processing for pavement distress analyses. In order to overcome the cost limitations in pavement data collection, this chapter presents a pervasive and 'smart' nature of the low-cost consumer-grade devices, in the acquisition of roadway condition data. By using such devices, no dedicated and expensive platforms and drivers are needed for automated data collection, and are as such suitable in the long-term in terms of costs, implementation and operations for road condition surveys.

Besides the data acquisition systems, in order to enhance the automation of pavement condition monitoring, there have also been advancements in the data collection techniques (e.g., [4–7]), and automated data processing techniques [8–10]. Because of the irregularities in terms of noise and topographic structure of pavement surfaces, more research is still ongoing on the accurate detection, classification and quantification of cracks and potholes. In addition, the computational costs for automated pavement distress detections are expensive, and better approaches are still necessary in the evaluation of the automated crack measurement systems under the various conditions [11].

The commercially available state-of-the-art systems, which comprise of digital camera and laser-illumination module, and laser road-imaging vehicles costs about $150,000. On the other hand, the pavement-surface profiler laser sensors, which are commonly used for measurement of road rutting-depth or surface-roughness, cost in the range of $130,000–$150,000. Comparatively, mobile pavement imaging techniques and manual inspection approaches respectively costs $88.5/mile and $428.8/mile, and the cost of using multi-sensor hybrid systems can range from $541/mile to $933/mile [2]. For fully automated pavement mapping systems, the cost of the imaging sensors and operations defines the purchase pricing, which averages at approximately $697,152 [12]. This chapter presents an approach for the customization of a low-cost imaging system, Kinect v2.0 sensor, as a prototype for cost-effective pavement imaging, and a data processing pipeline for pothole detection and extraction on asphalt pavements.

2. Measurement principle of the Kinect v2.0 RGB-D sensor

The Kinect v2.0 is the successor of the Xtion Pro Live RGB-D camera, called the Kinect v1.0. The version 2.0 Kinect RGB-D camera consists of a color (RGB) camera, an IR illuminator or projector and IR camera (**Figure 1(a)**). While the RGB camera records color information in high definition (HD), the IR projector emits an infrared laser and the IR camera is the sensor for the infrared laser. The Kinect v2 field in the horizontal is 70.6° and 60° in the vertical as depicted in **Figure 1(c)**. The values in the z-direction (depth values), are calculated using the Time of Flight (ToF) principle [16, 17], as shown in Eq. (1), and the x and y values are determined by using the homogeneous image coordinates u and v, and calculated as in Eqs. (2) and (3) [18]. The RGB and IR images acquired with the Kinect v2.0 partially overlap, because the RGB color camera has a wider horizontal field of view (FOV), and IR camera has a larger vertical FOV [15].

$$z = h = \frac{c \cdot \Delta\varphi}{4\pi f} \tag{1}$$

$$x = \frac{u - C_x}{f_x} \tag{2}$$

$$y = \frac{v - C_x}{f_y} \tag{3}$$

where z is the depth measure in meters, $\Delta\varphi$ is the phase shift, c is the speed of light and f is the modulation frequency; x is the horizontal position, u is the vertical image coordinate, C_x is optical center in the X-direction and f_x is the focal length in the X-direction, and y is the vertical position, v is the horizontal image coordinate, C_y is optical center in the Y-direction and f_y is the focal length in the Y-direction. In **Figure 1(b)**, **P** is the measured point on object surface, E is the IR emitter C is the IR sensor, and h or z is the unknown distance of measured point from sensor origin.

For the Kinect v1.0 RGB-D camera, the IR camera analyses a fixed speckle pattern projected by the IR projector and computes depth values by triangulation. This pattern analysis is referred to as the structured light (SL) approach, whereby a memorized IR pattern stored in the RGB-D camera's computer architecture is

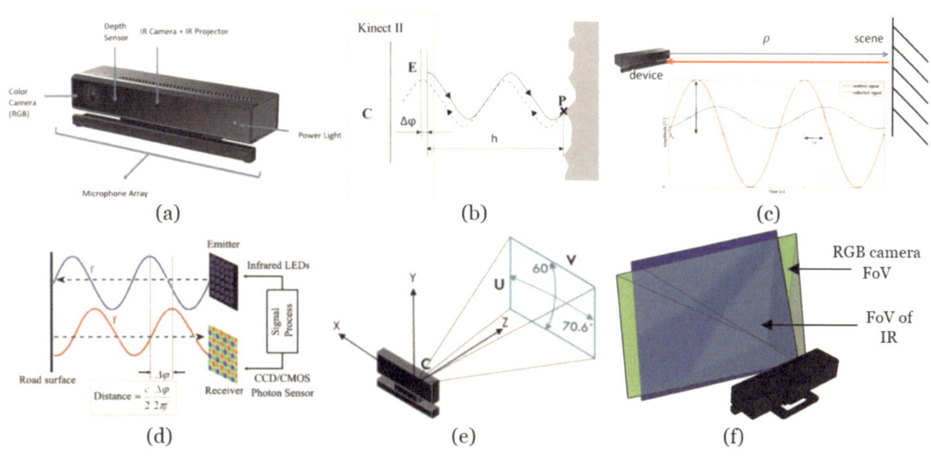

Figure 1.
(a) Kinect sensor v2.0 cameras; (b) and (c) principle of Time of Flight (ToF) phase measurement in Kinect v2.0, and (d) Kinect v2.0 and the field of view geometry [13, 14]. (e) Field of view (FoV) of Kinect v2.0 RGB and IR cameras [15].

projected onto the screen and compared with the current pattern on the screen [19]. If there are any obstacles in the way, the IR pattern changes shape from which the depth values can be deciphered. The Kinect v2.0 however, uses ToF technique to acquire depth values, where the sensor measures the time it takes for the modulated laser pulses from the IR projector to reach the object and then back to the IR camera [13]. The RGB resolution of the Kinect v2.0 is at 1920 × 1080 pixels, and the IR camera has a resolution of 512 × 424 pixels, with corresponding pixel sizes of 3.1 and 10 μm respectively. The collection of the (x,y,z) points results into 3D point

(a)

Specifications	Microsoft Kinect v1.0	SoftKinetic DS311	SoftKinetic DS325	SwissRanger SR4000
Range (short)	N/A	N/A	15 cm–1.5 m	N/A
Range (long)	0.8–4 m	1.5–4.5 m	N/A	0.8–8 m
Resolution (depth)	QVGA (640 × 480)	QVGA (320 × 240)	QVGA (160 × 120)	176 × 144
Field of view (H × V × D)	57.5° × 43.5° × N/A	57.3° × 42° × 73.8°	74° × 58° × 87°	43° × 34° × N/A
Technology (depth sensor)	Light coding	Depth sense	CAPD ToF	Time of Flight (ToF)
Frame rate (depth sensor)	30	25–60	25–60	50
Resolution (RGB)	640 × 480 or 1280 × 960	640 × 480	1280 × 720 (HD)	N/A
Field of view (RGB)	57.3° × 42° × N/A	50° × 40° × 60°	63.2° × 49.3° × 75.2°	N/A
Frame rate (RGB)	30	<25	<25	N/A
Power/data connection	USB 2.0 (1)	USB 2.0 (1)	USB 2.0 (1)	Lumberg M8 Male 3-pin
Size (W × H × D)	27.94 × 7.62 × 7.62 cm	24 × 5.8 × 4 cm	10.5 × 3.1 × 2.7 cm	6.5 × 6.5 × 6.8 cm
Price	$99	$299	$249	$4295

(b)

Parameter specification	Kinect v1.0	Kinect v2.0
Resolution of RGB camera (pixel)	640 × 480 or 1280 × 960	1920 × 1080
Resolution of IR and depth camera (pixel)	640 × 480	512 × 424
Field of view (FOV) of color camera	62° × 48.6°	84.1° × 53.8°
Field of view (FOV) of IR and depth image	57.5° × 43.5°	70.6° × 60°
Tilt motor	Yes	No
Maximum skeletal tracking	2	6
Method of depth measurement	Structured light	Time-of-Flight (ToF)
Depth distance working range	0.8–4.0 m	0.5–4.5 m
USB	2.0	3.0
Price	$99	$200

Table 1.
Comparative specifications of Kinect v1.0 and Kinect v2.0 and other low-cost sensors.

DOI: http://dx.doi.org/10.5772/intechopen.88877

cloud. This implies at the acquisition rate of 30 frames per second (fps), every frame of the Kinect v2.0 outputs 217,088 colored 3D points. The advantage that the Kinect v2.0 has over its predecessor Xtion Pro Live (Kinect v1.0), is that since it uses the principle of the ToF instead of relying on projected IR patterns for computing depth, the interference problem is greatly reduced as the sensor does not have to compute distances between neighboring points on the pattern [13]. The other advantage with the Kinect v2.0 over the Xtion, is that the camera has a built in ambient-light rejection method, which makes it possible to use in an outdoor environment with near infrared sources of interference [16]. **Table 1(a)** presents a summary of the differences between the Microsoft Kinect sensor v1.0 and other low-cost sensors, and **Table 1(b)** presents the fundamental characteristics of the Kinect versions 1.0 and 2.0.

3. Low-cost hardware system design and set-up for pavement data acquisition using Kinect v2.0

The establishment and design of an optimal low-cost imaging system, comprising of the hardware platform and peripheral requirements, with interface for Kinect-computer data acquisition, visualization and storage, in both static and dynamic acquisition modes is illustrated in **Figure 2**, and is termed as integrated Mobile Mapping Sensor System (iMMSS). For the implementation of the iMMSS, two main sets of equipment are used: (i) the Kinect v2.0—for RGB, Infrared (IR) and depth data capture, and (ii) a DC-AC power inverter—12 V DC to AC 220 V/200 W output. The power inverter is adaptable to the car charger port for powering the Kinect sensor for static and continuous pavement data acquisition modes. The iMMSS data acquisition system hardware-software set-up is as illustrated in the photo in **Figure 2**. The three main criteria in the field experimentation using the iMMSS comprise of: the shooting angle (vertical and oblique), shooting distance from the pavement, and the overall target positioning. **Figure 2** illustrates the hardware layout and software data capture system. The sensing device is housed within a sensor rack mounted onto the exterior of the wagon. To improve the contrast of the Kinect's laser pattern over the road surfaces, from the reflected IR radiation from sunlight an umbrella was used to block the rays from the sun and to create a shadow.

In terms of data acquisition in static and dynamic mode (**Figure 2**), the Kinect sensor captures depth and color images simultaneously at a frame rate of up to 30 fps. The integration of depth and color data results in a colored point cloud that

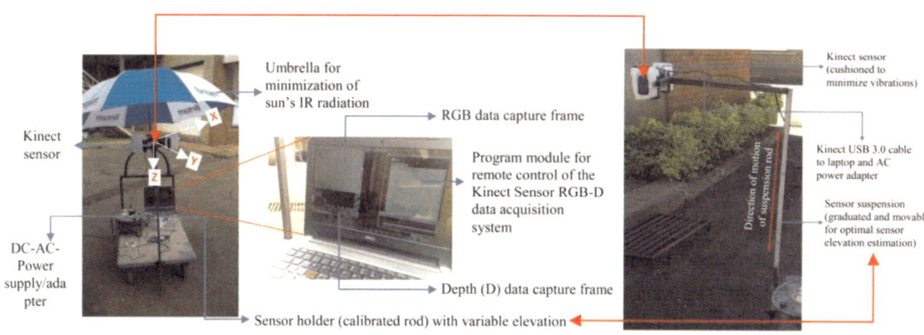

Figure 2.
iMMSS hardware-software set-up for road pavement data capture, visualization and storage using the Kinect sensor.

contains about 300,000 points in every frame. By registering the consecutive depth images it is possible to obtain an increased point density, and to create a complete point cloud. To realize the full potential of the sensor for mapping applications an analysis of the systematic and random errors of the data is necessary. The correction of systematic errors is a prerequisite for the alignment of the depth and color data, and relies on the identification of the mathematical model of depth measurement and the calibration parameters involved. The characterization of random errors is important and useful in further processing of the depth data, for example in weighting the point pairs or planes in the registration algorithm [20].

1. Pothole detection and the bias field effect

Under perfect conditions, potholes tend to have two visual properties characterized by: (i) low-intensity areas that are darker than nearby pavement because of road surface irregularity [21], and (ii) the texture inside the potholes being coarser than the nearby pavement [1, 22]. However, as illustrated in [8, 23], the pothole area is not always darker than nearby pavement. Furthermore, the irregularity of the road surface produces shadows at pothole boundaries, which is darker than nearby pavement. These conditions results into the lower accuracy of pothole detection using visual 2D techniques as was reported in [8]. In RGB imagery, pothole detection is influenced by the spill-in and spill-out phenomenon [1, 8], which is typically characterized by the similarities in the defect and non-defect features and regions. These results in the corruption of the defect regions on the pavement, with a smoothly varying intensity inhomogeneity called bias field. Bias is inherent to pavement imaging, and is associated with the imaging equipment limitations and also the pavement surface noise [1, 2].

Bias field in pothole detection can be modeled as a multiplicative component of an observed image, and varies spatially because of inhomogeneities, and can be modeled as in Eq. (4).

$$Y_j = B_j X_j + n \quad (4)$$

where Y_j is the measured image at voxel j; X_j is the true image signal to be restored; B_j is an unknown noise or bias field, and n is the additive zero-mean Gaussian noise. Eq. (4) modeled as an additive component by applying a logarithmic transformation, it is possible to obtain a simplified form as:

$$y_j = x_j + b_j \quad (5)$$

where x_j and y_j are the true and observed log transformed intensities at the jth voxel, respectively, and b_j is the noise or bias field at the jth voxel.

Bias or noise can be corrected by using prospective and retrospective methods. Prospective methods for noise minimization aim at avoiding the intensity inhomogeneities in the image acquisition process. Prospective methods are capable of correcting intensity inhomogeneity induced by the imaging devices; they are not able to remove object-induced effects. Retrospective methods in contrast, rely only on the information in the acquired images, and can thus remove intensity inhomogeneities regardless of their sources. The obvious choice in noise minimization is therefore the retrospective methods, which include filtering, surface fitting, histogram, and segmentation. Among the retrospective methods, segmentation-based approaches are particularly attractive, as they unify the tasks of segmentation and bias correction into a single framework. When an observed pixel y_j is defined as noisy, the neighboring pixels can be used to correct it, since the pixel is expected to be similar to its surrounding pixel. That is, the data points with similar feature vectors can be grouped into a single cluster and the data points with dissimilar feature vectors are also grouped into different clusters. By using a pre-segmentation clustering algorithm, the Euclidean distance between neighboring pixels is computed and used for the *a priori* clustering. This means that pixels that produce the lowest distance values to their neighbors are categorized as being nearly similar. Two pixels with similar neighboring values are expected to be close to each other, and hence the pixels can be clustered together. On way of minimizing noise through clustering is by using the *k*-means clustering algorithm, whereby the distance measure between every point $z_j^{(j)}$, and the cluster v_j is optimized by calculating the Euclidean distance measure $\left\|z_i^{(j)} - v_j\right\|^2$. The value of this distance measure function is an indicator of the proximity of the n data points to their cluster prototypes. Once the pre-clustering is carried out, a more robust segmentation approach can then be applied, to cluster the smoothened pavement image.

Image segmentation can be performed using different techniques such as: thresholding, clustering, transform and texture based methods [24]. Histogram-based thresholding is the simplest and often used approach [25]. Many global and local thresholding methods have been developed. While the global thresholds segment the entire image, with a single threshold using the gray-level histogram, the local based thresholds partition the image into a number of sub-images and select a threshold for each of the sub-image. The global thresholding methods select the thresholding based on different criterion such as: Otsu's method [24], minimum error thresholding [26], and entropic method [27]. These one-dimensional (1D) histogram thresholding methods work well when the two consecutive gray levels of the images are distinct. Further, all the 1D thresholding techniques do not combine the spatial information and the gray-level information of the pixels into the segmentation process. The performance of the thresholding techniques will lead to misclassifications in inherently correlated imagery, which are already corrupted by noise and other artifacts.

Real-world images are often ambiguous, with indistinguishable histograms. As such, it is complicated for the classical thresholding techniques to find criterion of similarity or closeness for optimal thresholding. This ambiguity in image segmentation can be solved by using fuzzy set theory, as a probabilistic global image segmentation approach. Using the conventional FCM formulation, each class is assumed to have a uniform value as given by its centroid. Similarly, each data point is also assumed to be independent of every other data point and spatial interaction between data points is not considered. However, for image data, there is strong correlation between neighboring pixels. In addition, due to the intensity non-uniformity artifacts, the data in a class no longer have a uniform value. Thus to realize meaningful segmentation results, the conventional FCM algorithm has to be modified to take into account both local spatial continuity between neighboring data and intensity non-uniformity artifact compensation. This chapter illustrates the use of spatial fuzzy *c*-means (SFCM), so as to incorporate the spatial neighboring information into the standard fuzzy *c*-means for pothole detection on pavement surfaces.

3.1 Fuzzy *c*-means clustering with spatial constraints

FCM is an unsupervised fuzzy clustering algorithm. The conventional clustering algorithms determine the "hard partition" of a given dataset based on certain criteria that evaluates the goodness of partition, so that each datum belongs to exactly one cluster of the partition. The soft clustering on the other hand finds the "soft partition" of a given dataset. And in "soft partition", the datum can partially belong to multiple clusters. Soft clustering algorithms do generate a soft partition that also forms fuzzy partition. A type of soft clustering of special interest is one that ensures membership degree of point x_j in all clusters adding up to one (Eq. (6)), and also satisfies the constrained soft partition condition.

$$\sum_i \mu_{c_i}(x_j) = 1, \forall x_j \in X \tag{6}$$

The fuzzy c-means is a clustering method which allows one piece of data to belong to two or more clusters [28, 29]. The standard FCM algorithm considers the clustering as an optimization problem where an objective function must be minimized, and assigns pixels to each category by using fuzzy memberships. If $I = \{x_j \in R^d\}_{j=1,\dots,N}$ is a $p \times N$ data matrix, where, p represents the dimension of each x_j "feature" vector, and N represents the number of feature vectors (pixel numbers in the image), then the FCM algorithm is an iterative optimization that iteratively minimizes the objective function, with respect to fuzzy membership $'U'$, and set of cluster centroids, $'V'$as in Eq. (7).

$$J_{FCM} = \sum_{j=1}^{N} \sum_{i=1}^{c} u_{ij}^m \cdot \|x_j - v_i\|^2 \tag{7}$$

where u_{ij} represents the fuzzy membership of pixel x_j in the ith cluster and $u = (u_1, u_2, \dots, u_c)$ are the set of cluster centers; $'C'$ is the number of clusters; $v_i$is the ith cluster center; $\|\cdot\|$ is a Euclidean distance or the norm metric, and m is a constant for fuzziness exponent. The parameter m controls the fuzziness of the resulting partition or the fuzziness of the consequential partition, and $m = 2$ is used in this study.

The cost function is minimized when pixels close to the centroid of their clusters are assigned high membership values, and low membership values are assigned to pixels with data far from the centroid. The membership function represents the probability that a pixel belongs to a specific cluster. In the FCM algorithm, the probability is dependent solely on the distance between the pixel and each individual cluster center in the feature domain. By minimizing Eq. (7) using the first derivatives with respect to u_{ij} and v_i then setting them to zero using the Lagrange method, the membership functions and cluster centers are updated by solutions of u_{ij} and the fuzzy centers v_i:

$$u_{ij} = \frac{1}{\sum_{k=1}^{c} \left(\frac{\|x_j - v_i\|}{\|x_j - v_k\|} \right)^{2/(m-1)}} \tag{8}$$

and

$$v_i = \frac{\sum_{j=1}^{N} u_{ij}^m x_j}{\sum_{j=1}^{N} u_{ij}^m} \tag{9}$$

Starting with an initial guess for each cluster center, the FCM converges to a solution for v_i representing the local minimum or a saddle point of the cost function. Convergence can be detected by comparing the changes in the membership function or the cluster center at two successive iteration steps. In an image, as illustrated in [1], the neighboring pixels are normally highly correlated. This is because these neighboring pixels possess similar feature values, and the probability that they belong to the same cluster is often high. The introduction of the spatial information is an important cue in resolving the *mixel* problem within a pavement pothole *voxel*. While this spatial relationship is important in clustering, it is not utilized in a standard FCM algorithm. To overcome the effect of noise in the segmentation process, [30] proposed spatial FCM algorithm in which spatial information can be incorporated into fuzzy membership functions directly using a spatial function. The spatial information is introduced while updating the membership function u_{ij} in the repetitive FCM algorithm because the neighborhood pixels possess same properties as the center pixel. To exploit the spatial information, the spatial function is defined by h_{ij} (Eq. (10)).

$$h_{ij} = \sum_{k \in NB(x_j)} u_{ik} \tag{10}$$

where $NB(x_j)$ is a local square window centered on pixel x_j in the spatial domain, and in this illustration, a 5×5 window is used.

Like the membership function, the spatial function h_{ij} represents the probability that pixel x_j belongs to the ith cluster. The spatial function of a pixel for a cluster is large if the majority of its neighborhood belongs to the same clusters. The spatial function is used in updating the membership function again, and is incorporated into membership function as follows as presented in Eq. (11) [30].

$$u'_{ij} = \frac{u_{ij}^p h_{ij}^q}{\sum_{k=1}^{c} u_{kj}^p h_{kj}^q} \tag{11}$$

where p and q are two parameters used to control the relative importance of both the membership and spatial functions respectively.

In a homogenous region within an image, the spatial functions will strengthen the original membership, and the clustering result remains unchanged. However, for a noisy pixel, this formula reduces the weighting of a noisy cluster by the labels of its neighboring pixels. As a result, misclassified pixels from noisy regions or spurious blobs can easily be corrected. The spatial FCM with parameter p and q is denoted $SFCM_{p,q}$. For $p = 1$ and $q = 0$, the $SFCM_{1,0}$ is identical to the conventional or standard FCM. In the $SFCM_{p,q}$, the objective function is not changed, instead the membership function is updated twice. The first update is the same as in standard FCM that calculates the membership function in the spectral domain. However in the second phase, the membership information of each pixel is mapped to the spatial domain, and the spatial function is computed from that. The spatial function is defined as the sum of the membership values in spatial domain in the entire neighborhood around the pixel under consideration. The FCM iteration proceeds with the new membership that is incorporated with the spatial function. The iteration is stopped when the maximum difference between two cluster centers at two successive iterations is less than a threshold (=0.02). After the convergence, defuzzification is applied to assign each pixel to a specific cluster for which the membership is maximal. The $SFCM_{p,q}$ works well for high as well as low density noise, and can be applied for single and multiple feature data. As compared to other methods FCM based methods, $SFCM_{p,q}$ gives superior results without any boundary leakage even at high density noise, when the q value is carefully selected [31].

3.2 Depth image data smoothing and hole-filling

To correctly analyze and potentially combine the RGB image with the depth data, the spatial alignment of the RGB and the depth camera outputs is necessary. Additionally, the raw depth data are very noisy and many pixels in the image may have no depth due to multiple reflections, transparent objects or scattering in certain nearby surfaces. As such the inaccurate and or missing depth data (holes) need to be recovered prior to data processing. The recovery is conducted through application-specific camera recalibration and or depth data filtering. In this section we deal with the depth data filtering first, and in the next subsection, the camera calibration is discussed. By enhancing the depth image using color image, the following issues are addressed: (i) due to various environmental reasons, specular reflections, or simply the device range, there are regions of missing data in the depth map; (ii) the accuracy of the pixels values in the depth image is low, and the noise level is high. This is true mostly along depth edges and object boundaries, which is exactly where such information is most valuable; (iii) despite the calibration, the depth and color images are still not aligned well enough. They are acquired by two close, but not similar, sensors and may also have differences in their internal camera properties (e.g., focal length). This misalignment leads to small projection differences, even, again, these small errors are more noticeable especially along edges, and (iv) usually the depth image has lower resolution than the color image, and therefore it should be up-sampled in a consistent manner.

Because of the limitations in the depth measuring principle and object surface properties, the depth image from Kinect inevitably contains optical noise and unmatched edges, together with holes or invalid pixels, which makes it unsuitable for direct application [32]. In order to remove noise from the depth image, the joint bilateral filter is preferred. This is because the joint bilateral filter has the advantage of preserving edges while removing noises, analyzing through every image pixel and replacing every image pixel-by-pixel with the median of the pixels in the corresponding filter region R. This process can be expressed according to Eq. (12).

$$I'(u,v) \rightarrow median\{I(u+i, v+j)|(i,j) \in R\} \tag{12}$$

where, (u,v) is the position of the image pixel and (i,j) is the neighborhood size of the image region and these are specified as a two element numeric vector of positive integers. By using median filtering, each output pixel contains the median value in the $i \times j$ neighborhood around the corresponding pixel in the input image.

In filling holes in depth images: (i) [33] used bilateral filter and median filter in the temporal domain; (ii) [34] proposed joint bilateral filter and Kalman filter for depth map smoothing, and to reduce the random fluctuations in the time domain. Jung [35] proposed a modified version of the joint trilateral filter (JTF) by using both depth and color pixels to estimate a filter kernel and by assuming the presence of no holes. Liu et al. [36] employed an energy minimization method with a regularization term to fill the depth-holes and remove the noise in depth images. The linear regression model utilized was based on both depth values and pixel colors. From the above studies, it is noted that the methods are primarily based on different types of filters to smooth noise in depth images and to fill holes by using color images to guide the process.

Introduced by [37], the bilateral filter is a robust edge-preserving filter with two filter kernels: a spatial filter kernel and a range filter kernel, which are traditionally based on a Gaussian distribution, for measuring the spatial and range distance between the center pixel and its neighbors, respectively [38].

By letting I_X be the color at pixel **x**, and I_X^I be the filtered value, it is desired for I_X^I to be:

$$I_X^I = \frac{\sum_{y \in N(x)} f_S(x,y) \cdot f_R(I_x, I_y) \cdot I_y}{\sum_{y \in N(x)} f_S(x,y) \cdot f_R(I_x, I_y)} \tag{13}$$

where **y** is a pixel in the neighborhood $N(\mathbf{x})$ of pixel **x**, where $f_S(x,y) = \exp\left(-\frac{\|\mathbf{x}-\mathbf{y}\|^2}{2\sigma_S^2}\right)$ and, $f_R(I_x, I_y) = \exp\left(-\frac{\|I_x - I_y\|^2}{2\sigma_R^2}\right)$ are the spatial and range filter kernels measuring the spatial and range/color similarities. The parameter σ_S defines the size of the spatial neighborhood used to filter a pixel, and σ_R controls how much an adjacent pixel is down-weighted because of the color difference.

The limitation of the conventional bilateral filter is that it can interpret impulse noise spikes as forming an edge. A joint or cross bilateral filter [39, 40] is similar to the conventional bilateral filter except that in the case of the joint bilateral filter, the range filter kernel $f_R(\cdot)$ is computed from another image called the *guidance image*. The guide image J indicates where similar pixels are located in each neighborhood. With J as the guidance image, then the joint bilateral filtered value at pixel **x** is determined as in Eq. (14).

$$I_X^J = \frac{\sum_{y \in N(x)} f_S(x,y) f_R(J_x, J_y) I_y}{\sum_{y \in N(x)} f_S(x,y) f_R(J_x, J_y)} \tag{14}$$

It is important to note that the joint bilateral filter ensures the texture of the filtered image I^J to follow the texture of the guidance image J. In the implementation this paper, the image intensity was normalized such that it ranges from [0, 1], and image coordinates were also normalized so that **x** and **y** also reside in [0, 1].

With this depth hole filling based on the bilateral filter, the depth value at each pixel in an image is replaced by a weighted average of depth values from nearby pixels. While the joint bilateral filter has been demonstrated to be very effective for color image upsampling, if it is directly applied to a depth image with a registered RGB color image as the guidance image, the texture of the guidance image (that is independent of the depth information) is likely to be introduced to the upsampled depth image, and the upsampling errors mainly reside in the texture transferring property of the joint bilateral filter [38]. Meanwhile, the median filtering operation minimizes the sum of the absolute error of the given data [41], and is much more robust to outliers than the bilateral filter. A possible solution to the "hole-filling" problem in depth imagery is to focus on the combination of the median operation with the bilateral filter so that the texture influence can be better suppressed while maintaining the edge-preserving property [42].

3.3 Calibration of RGB and IR Kinect cameras

Despite the fact that the Kinect, like other off-the-shelf sensors, has been calibrated during manufacturing, and the camera parameters are stored in the device's memory, this calibration information not accurate enough for reconstructing 3D information, from which a highly precise cloud of 3D points should be obtained. Furthermore, the manufacturer's calibration does not correct the depth distortion, and is thus incapable of recovering the missing depth [43]. Using a 9×8

checkerboard, with 30 mm square fields, a set of close-up RGB/IR images of the checkerboard placed in different positions and orientations (**Figure 3(a)**), can be collected and used for calibration. The Bouguet's Camera Calibration Toolbox [44] in MATLAB can be used for the identification of RGB and IR camera parameters, utilizing the two versions of Herrera's method [45]. IR camera calibration, the IR emitter should be disabled during imaging so as to achieve appropriate light conditions. The output matrices for the intrinsic, distortion and extrinsic calibration parameters are presented in **Table 2**.

3.3.1 Initialization of intrinsic and extrinsic calibration

For the color camera, the initial estimation of I_c and $T_c^{(i)}$ for all calibration images is carried out as described in Bouguet's toolbox. The intrinsic parameters for the depth camera are defined as $I_d^{'} = \{\mathrm{f}_d, \mathrm{c}_d, \mathrm{k}_d, c_0, c_1\}$, since the depth distortion terms are not considered. They are initialized using preset values, which are publicly

Figure 3.
Checkerboard RGB (top) images and the corresponding IR (bottom) calibration images. From the case study roads, a database of 10,540 color and depth test image frames has been acquired and being processed.

	Intrinsic calibration matrix		
536.782668	0.000000	319.133028	
0.000000	536.889190	258.356500	
0.000000	0.000000	1.000000	
	Distortion calibration matrix		
0.243645	−0.572745	−0.008210	0.000119
	Extrinsic calibration matrix		
0.999987	−0.004894	−0.001283	110.506445
−0.004661	−0.989735	0.142836	−133.830468
−0.001969	−0.142828	−0.989746	867.124291
0.000000	0.000000	0.000000	1.000000

Table 2.
Intrinsic, distortion and extrinsic calibration matrix parameters.

available for the Kinect, online. For each input disparity map i, the plane corners are extracted, defining a polygon. For each point x_d inside the polygon, the corresponding disparity d is used for computing a depth value z_d using $z = \frac{1}{c_1 d_u + c_0}$, where $d = d_u$ since the measured disparities are used, and c_0 and c_1 are part of the depth camera's intrinsics. The correspondences (x_d, y_d, z_d) are used for computing 3D X_c points originating a 3D point cloud. To each 3D point cloud, a plane is fitted using a standard total least squares algorithm.

3.4 Pothole search engine

As a pre-processing step and prior to the segmentation and clustering of the RGB and depth data, pothole search engine (PSE) is necessary. It is then possible to extract potholes-only images for further autonomous processing. This can be accomplished by using a 2-class k-means clustering of the candidate RGB image frames, and is confirmed using ellipsoidal fitting on the classified binary image frame.

3.4.1 k-means clustering and edge ellipse fitting for pothole search

Since the data collected comprises of pothole and non-pothole pavement defect image frames, the first preprocessing step after the calibration is to eliminate the non-pothole images from the database. Using unsupervised classification on the acquired RGB data frames, images with potential potholes are selected based on k-means clustering [46], and adaptive median filtering. From the candidate potholes images, edge lines are estimated and the corresponding ellipse(s) are fitted using least squares optimization. This algorithm is applied in a batch processing mode, and the efficiency of the approach is then confirmed by using visual inspection and comparison.

3.4.2 Horizontal and vertical integral projection (HVIP)

Integral projection (IP) has the discriminative to accumulate and resolve the pixel histograms into pothole and non-potholes pixels, by analyzing the horizontal and vertical (HV) pixel distributions within an image, represented by horizontal and vertical projections. Given a grayscale image $I(x, y)$, the horizontal and vertical IPs are defined as follows in Eqs. (15) and (16).

$$\mathrm{HP}\,(y) = \sum_{i \in x_y} I(i,j) \tag{15}$$

$$\mathrm{VP}\,(x) = \sum_{j \in y_x} I(i,y) \tag{16}$$

where HP and VP are the horizontal and vertical IP, respectively. x_y and y_x denote the set of horizontal pixels at the vertical pixel y and the set of vertical pixels at the horizontal pixel x, respectively.

3.4.3 Database search for candidate pothole image frames using ellipse fitting and HVIP

With a visual comparison of 99% efficiency for the pothole database search, **Table 3** shows the results using the pothole search engine (PSE). The ellipse detection indicates the presence of defect or no-defect within the image, and also defines the orientation of the pothole with respect to the longitudinal profile of the road.

Test data	Pavement RGB image frame	2-class *k*-means classification and adaptive median-filtering	Fitted edge on candidate pothole image	Ellipse-fitting in PSE for pothole presence detection	Horizontal and vertical integral projection (HVIP) plots
#1	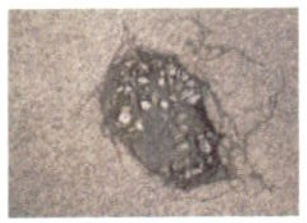	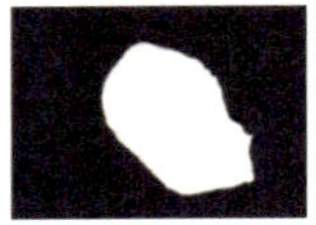			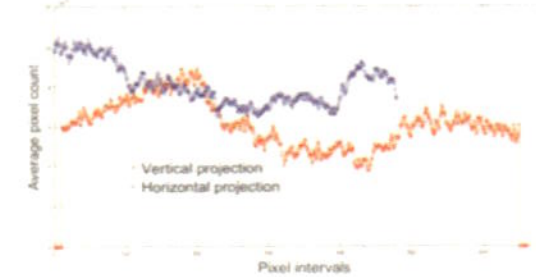
#2	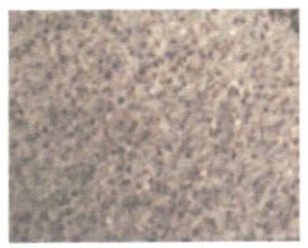			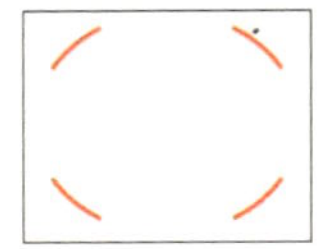	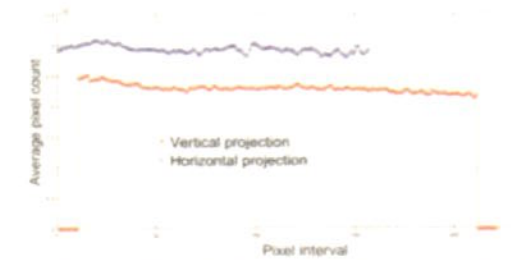
#3		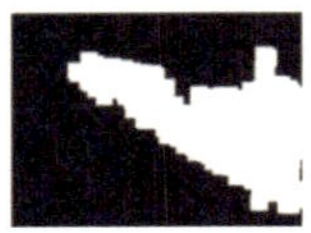	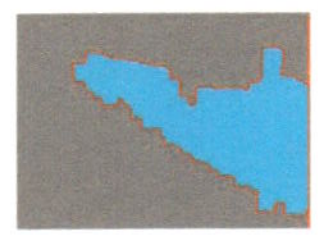	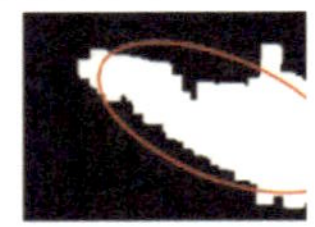	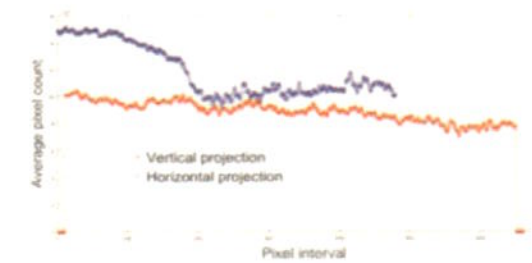
#4		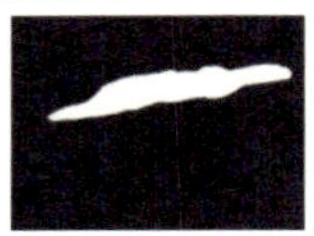	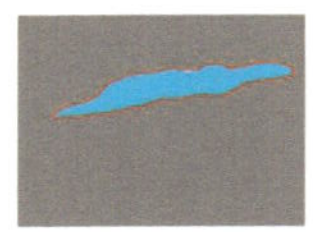	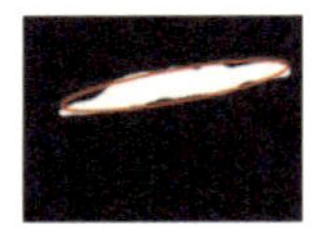	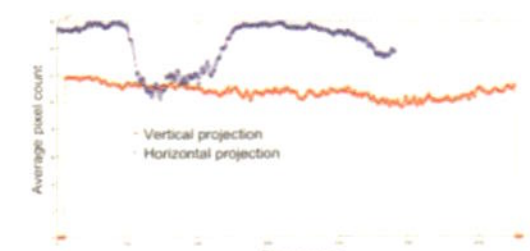

Table 3.
PSE and HV-integral projection search for pothole and non-pothole frames from RGB test data.

The results of horizontal and vertical IP (HVIP) analysis for several pavement images with varied sized pixels are presented in **Table 3**. As observed from the test results, a structurally healthy pavement image with non-potholes (e.g., test image #2) is generally characterized by recognizably stable signals of both horizontal and vertical integral projections. On the other hand, the integral projections of images containing potholes (e.g., test images #1, #3 and #4), has peak(s) in either the vertical or horizontal or both IPs, depending on the strength or the severity of the pothole and lighting conditions. Where both the horizontal and vertical signals are strong, the locations of the two peaks tend to be relatively close to each other. Thus in addition to the ellipsoidal fitting, HVIP can effectively be used in the extraction of pothole and non-pothole image frames in a pothole database search engine system. In the PSE search system, data acquired under varied illumination conditions were tested, to ensure the effectiveness of the system with data of different resolutions.

4. Pothole metrology data parametrization

Figure 4 illustrates the conceptual approximation of a pothole with dimensional parameters that define the pothole metrology as: width, depth, surface area and volume. Assuming the potholes have the shape of a circular paraboloid, then in 2D they can be represented by the function $f(x,y) = x^2 + y^2$.

4.1 Pothole depth determination using depth image

The depth-image plane (**Figure 4**) is one of the noise factors, whereby the plane is not necessarily parallel to the pavement surface. The noise points, which are the non-defect points between the pavement-pothole plane and the camera, have to be filtered out for the accurate depth detection and the subsequent 2D-pothole detection from the depth image. The general principle of removing the outlier points (noise), is by determining the local minimum of each column and then subtracting from the column itself in order to extract the pothole from the rest of data [47]. The minimum of each column defines the depth below which the pothole starts on the road pavement surface, and is referred to as the depth-image plane. Using this approach, the depths d_i including the maximum depth $d_{i_{max}}$ can be quantified, and the mean depth $\overline{d_i}$ for a given pothole is also computed.

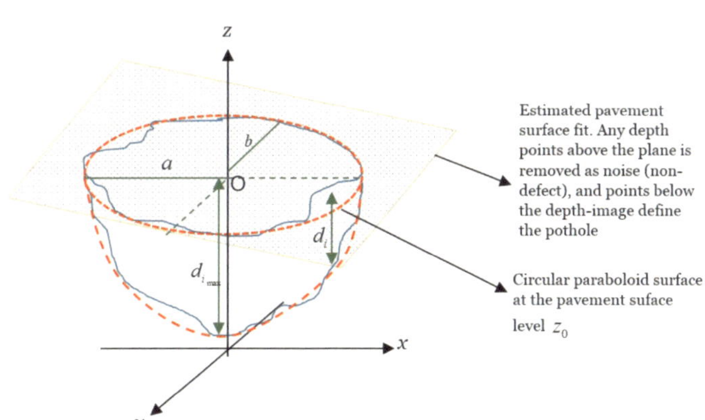

Figure 4.
Representation and approximation of pothole metrology elements: depth, width, surface area and volume.

4.2 Pothole width measurement

The width of a pothole can be defined by the semi-major a and semi-minor b axes, on the assumption that an ellipse, based on the major path elliptic regression, is used pothole shape extraction [48]. To determine the lateral width of the pothole, it can be estimated using a circular paraboloid, which is an elliptical paraboloid. And, an elliptical paraboloid is a surface with parabolic cross-sections in 2-orthogonal directions and 1-elliptical cross-section in the other orthogonal direction. Using an edge detection algorithm, the near-true shape of the pothole is first derived using the proposed SFCM, and then an elliptical fit is used to approximate the shape, from which the axes are defined for the calculation of the surface area and volume of the pothole.

4.3 Pothole surface area determination

In order to determine the surface area of the pothole, the optimally detected edge is used to fit the shape of the pothole as either elliptic paraboloid or circular paraboloid. While the former is defined by the dimensions of semi-major axis a and semi-minor axis b, the latter is defined by the estimated radius r. The surface area is then computed by using the surface integrals of either of the paraboloids [49], as respectively shown in Eqs. (17) and (18) for the elliptic and circular paraboloids.

$$A = \pi ab \tag{17}$$

$$A_r = \frac{\pi}{6}\sqrt{(1 + 4r^2)^3} \tag{18}$$

If pixels counts are used, then Eq. (19) can be implemented, [8]. Whereby in Eq. (19), l is the pixel size and I_p is the binary value of pixel at coordinate position (x,y). The area A_p is estimated on the basis of the average of a 2×2 window.

$$A_p = l^2 \cdot \sum_x \sum_y I_p(x, y) \tag{19}$$

4.4 Pothole volume estimation

According to [50], if T is a closed region bounded by a surface S, and F is a vector field defined at each point of T and on its boundary surface, then $\int\int\int_T F dv$ is the volume integral of F through the bounded region T. As in case for the surface area of a pothole, the area is either estimated by an elliptic paraboloid or a circular paraboloid. The volume of the elliptic paraboloid V can be estimated according to Eq. (20), and the volume V_r Of the pothole is estimated using a circular paraboloid as in Eq. (21).

$$V = \frac{4}{3}\pi abd_{\max} \tag{20}$$

$$V_r = \frac{\pi r^4}{2} \tag{21}$$

Since the depth for each pixel d_i is obtainable from the depth image, the integration of all small volumes represented by each pixel leads to the total volume of area within the frame [51]. Therefore the estimated volume V_d in terms of the pixel depth is given by Eq. (22)

$$V_d = l_p^2 \cdot \sum_y \sum_x I_d(x, y) \cdot I_p(x, y) \tag{22}$$

where V_d is the total pothole volume, and $I_d(x, y)$ is depth of pixel p at location (x, y).

4.5 Prototype implementation strategy for pothole detection using low-cost sensor

Figure 5 illustrates the processing steps in implementing the detection, and visualization potholes and related metrological parameters from the Kinect v2.0 RGB-D, based on the experimental iMMSS data capture system. In summary the processing system should comprise of data acquisition and geometric transformation; preprocessing for noise minimization; cascaded pothole detection approach from fused RGB-D data using dual-clustering approach comprising of k-means and spatial fuzzy c-means, and a parallel processing system for pothole area and volume detection from RGB and depth imagery.

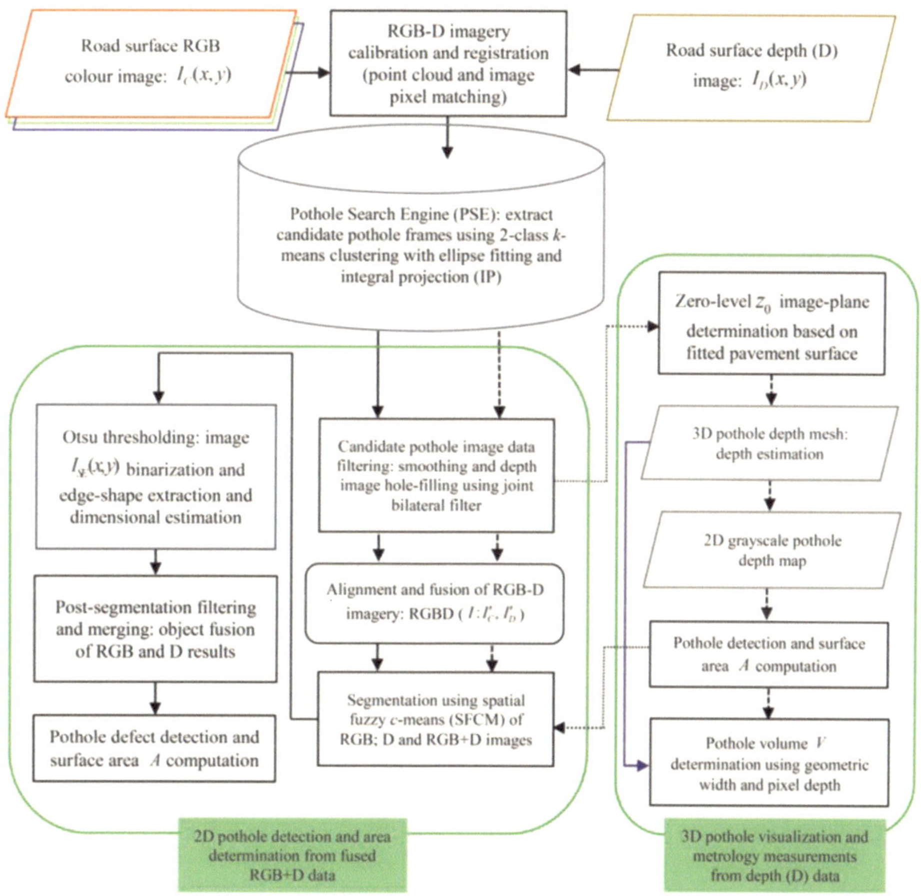

Figure 5.
Processing pipelines for pothole detection based on cascaded dual-clustering and pothole metrology quantification and visualization from multimodal iMMSS low-cost RGB-D sensor system.

5. Some experimental results and analysis

5.1 Pothole detection using SFCM segmentation

The results for the clustering of the RGB imagery using FCM and SFCM are comparatively presented. Where there is low spectral heterogeneity, the first Principal Components Transform image (PCT-band 1) is used in the FCM and SFCM clustering. The results in **Table 4** shows that the inclusion of the spatial neighborhood information using the SFCM, results in a more compact detection of the potholes, by segmenting the potholes from the non-potholes and ensuring homogeneity within the pothole itself, hence taking the spatial cues in clustering. Furthermore, the SFCM performs much better than FCM especially under different lighting conditions.

5.2 Pothole depth imagery representation

Defects on pavements are defined as surface deformations that are greater than a threshold as illustrated in **Figure 6(b)**. Since the captured depth data is corrupted with noise, the depth-image plane as illustrated in **Figure 4** (**Figures 6(b)** and **6 (c)**), is not necessarily parallel to the surface that is under inspection. This is solved by fitting a plane to the points in the depth image (**Figure 6(b)**), that are not farther than a threshold from the IR camera (**Figure 6(c)**). By using the random sample consensus (RANSAC) algorithm [52], the plane is fitted to the points, and the depth image is subtracted from the fitted plane, with the results in **Figure 6(d)**. To discriminate between the depressions (potholes) and the flat regions (non-potholes), the Otsu's thresholding algorithm is used. Sample results of the depth-image segmentation are sequentially presented in **Figure 6**.

5.3 Feature based RGB-D data fusion for enhanced pothole segmentation

In this section, an illustration on the potential of fusion of the depth and color image at the object or feature level is demonstrated. A possible two-way fusion approach comprising of either: (i) pre-pothole detection fusion involving the enhancement of the color image with the depth image, or (ii) post-pothole detection fusion of the pothole defect features as independently determined from the RGB and depth images respectively is proposed and conceptually represented in **Figure 7**. The first approach presents a joint segmentation approach, which is similar to extracting consistent layers from the image where each layer segment in terms of both color and depth. It is common for real scene object, like pavement pothole surfaces, to be characterized by different intensities and a small range of depths. The incorporation of the depth information into the segmentation process, allows for the detection of real pothole object boundaries instead of just coherent color regions, and the objective is to enhance the application relevant features in the resultant fused image product.

The potential and significance of fusion of RGB and depth imagery is illustrated in **Figures 8** and **9**, using the pothole edge identification from the RGB and depth image data. **Figure 8** shows an RGB and depth (RGB-D) single frame pavement data acquired Kinect experimental setup. The RGB is smoothened (left frame) using the median filter, while hole-filling using the joint bilateral filter is applied to the depth image (right frame). It is observed that the two images complement each other. Comparing the corrected image datasets, it is observed that the depth image clearly defines the pothole edges as compared to the fuzzy representation of the

Original RGB pothole image frame and the first principal components (PC) image	Clustering and segmentation results for pothole detection using fuzzy *c*-means clustering		
	FCM clustering	**SFCM segments and segmentation**	**SFCM results with Otsu thresholding**
	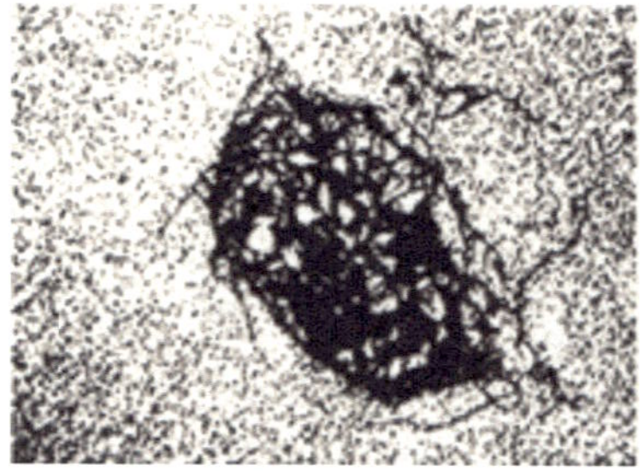 FCM clustering	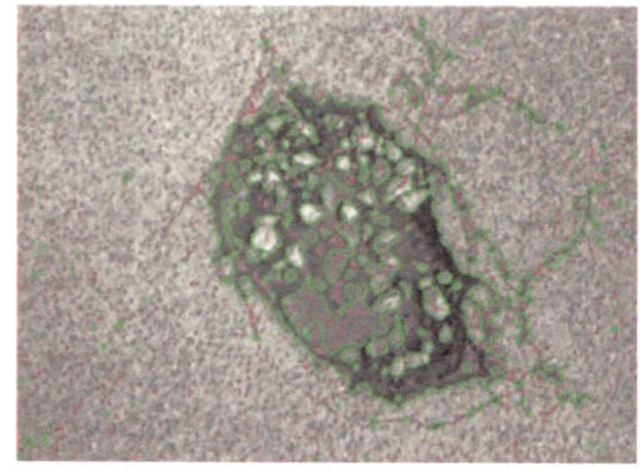 SFCM clustered segments	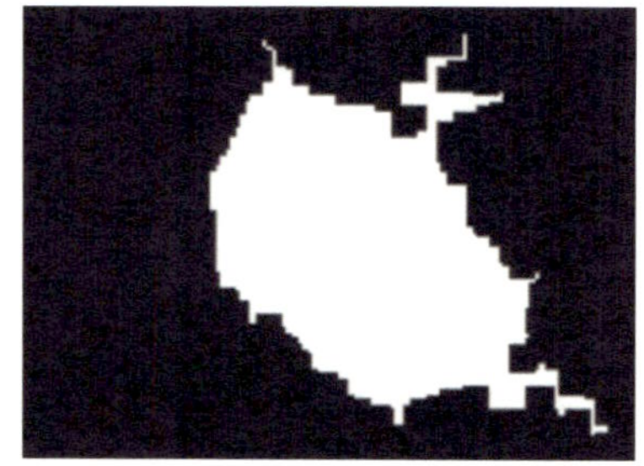 Detected pothole after thresholding
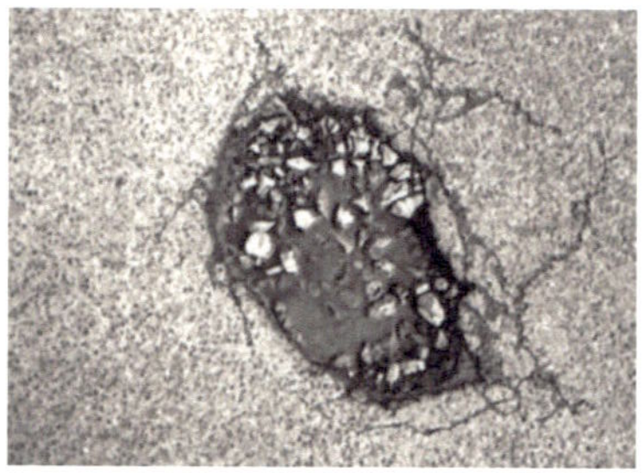	 FCM segmentation results with Otsu thresholding	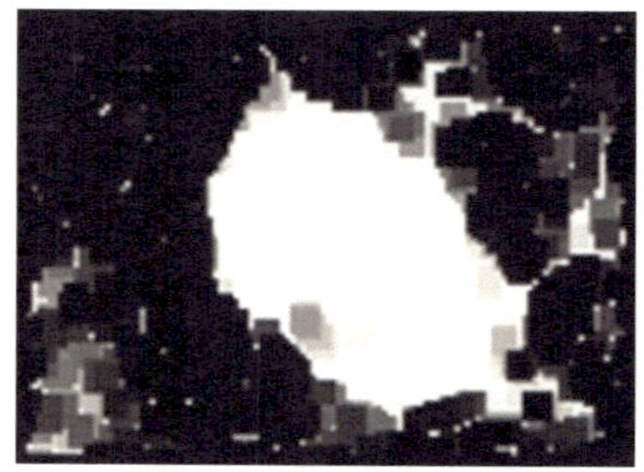 SFCM segmentation results	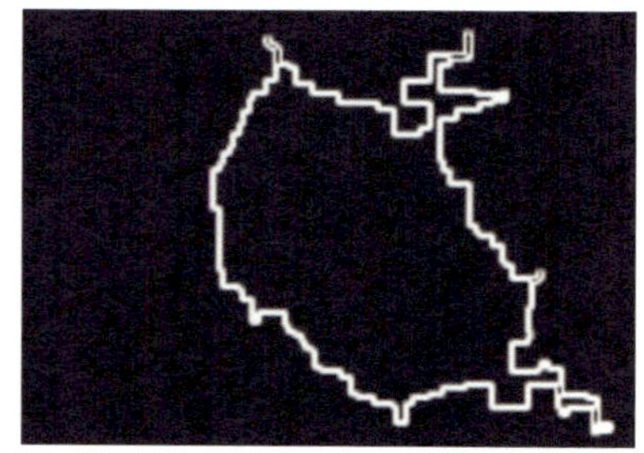 Perimeter of the detected pothole

Original RGB pothole image frame and the first principal components (PC) image	Clustering and segmentation results for pothole detection using fuzzy *c*-means clustering		
	FCM clustering	**SFCM segments and segmentation**	**SFCM results with Otsu thresholding**
	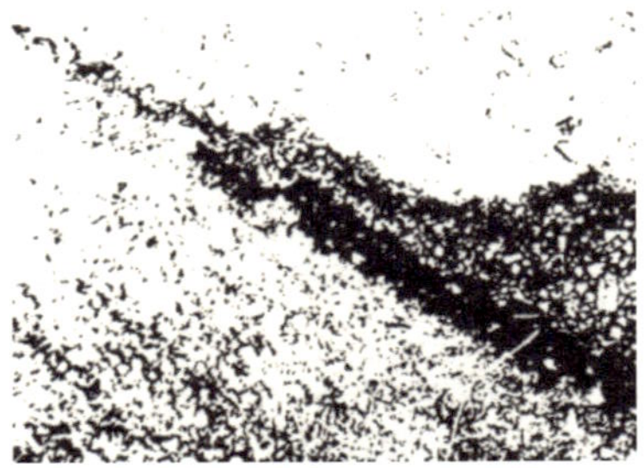 FCM clustering	 SFCM clustered segments	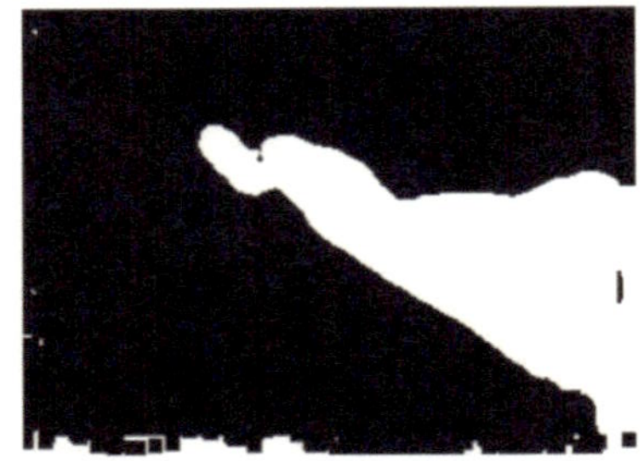 Detected pothole after thresholding
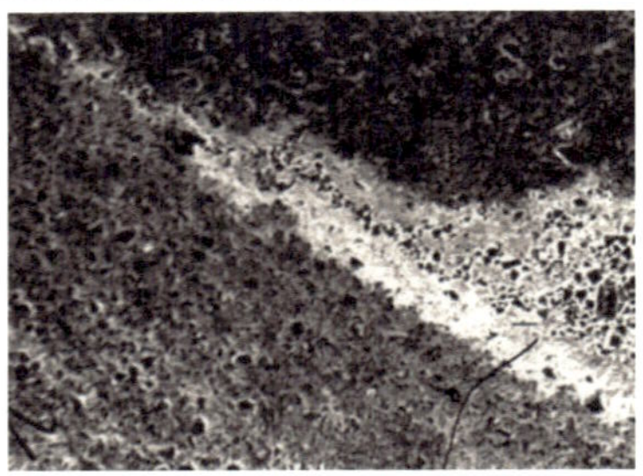	 FCM segmentation results with Otsu thresholding	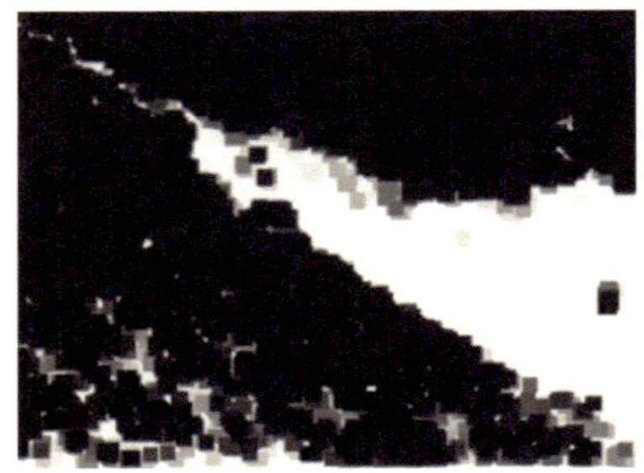 SFCM segmentation results	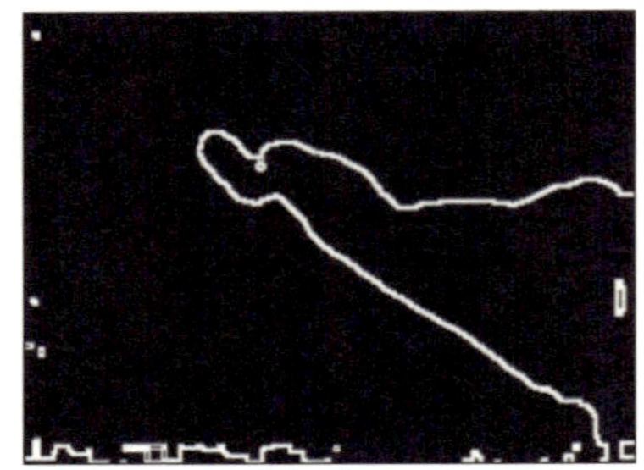 Perimeter of the detected pothole

DOI: http://dx.doi.org/10.5772/intechopen.88877

Original RGB pothole image frame and the first principal components (PC) image	Clustering and segmentation results for pothole detection using fuzzy *c*-means clustering		
	FCM clustering	**SFCM segments and segmentation**	**SFCM results with Otsu thresholding**
	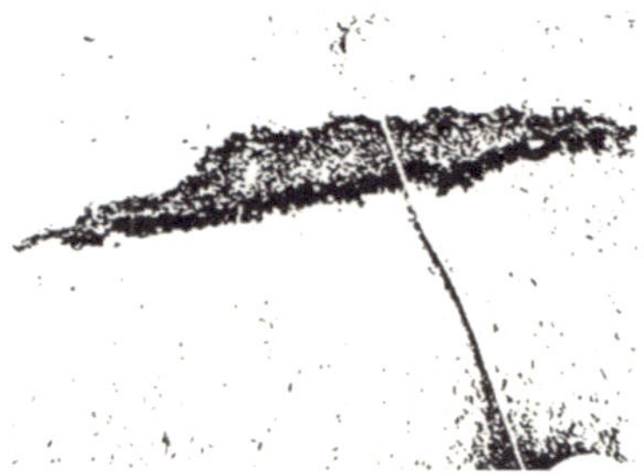 FCM clustering	 SFCM clustered segments	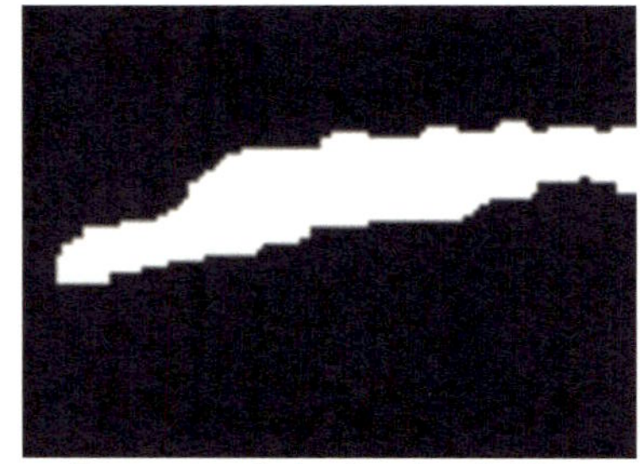 Detected pothole after thresholding
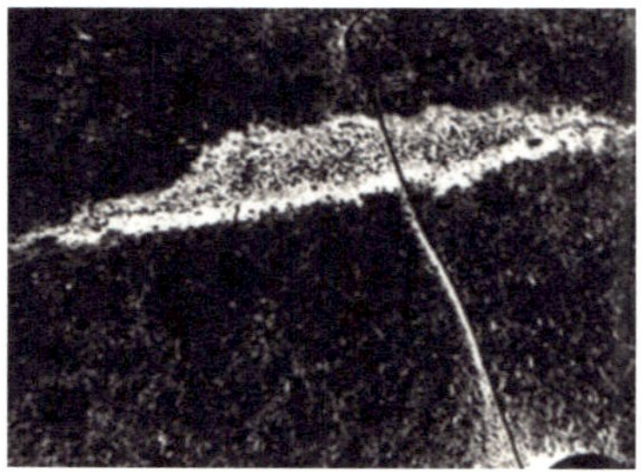	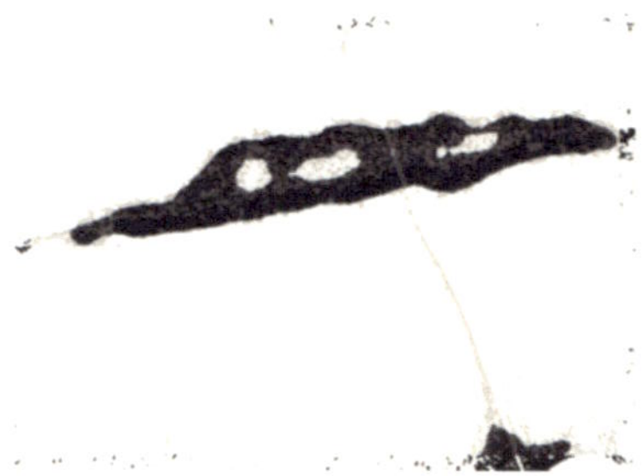 FCM segmentation results with Otsu thresholding	 SFCM segmentation results	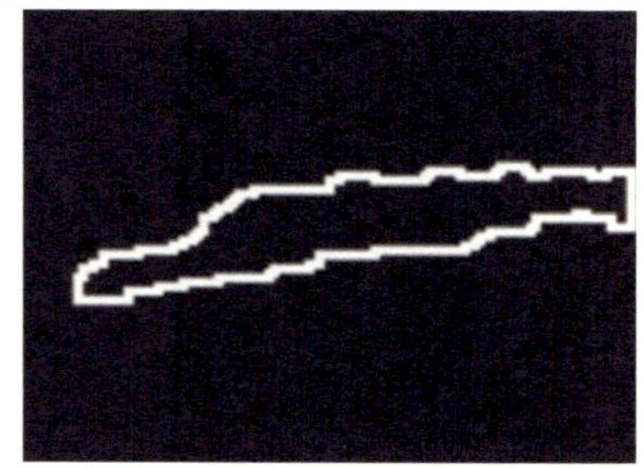 Perimeter of the detected pothole

Table 4.
Pothole detection results using FCM and SFCM.

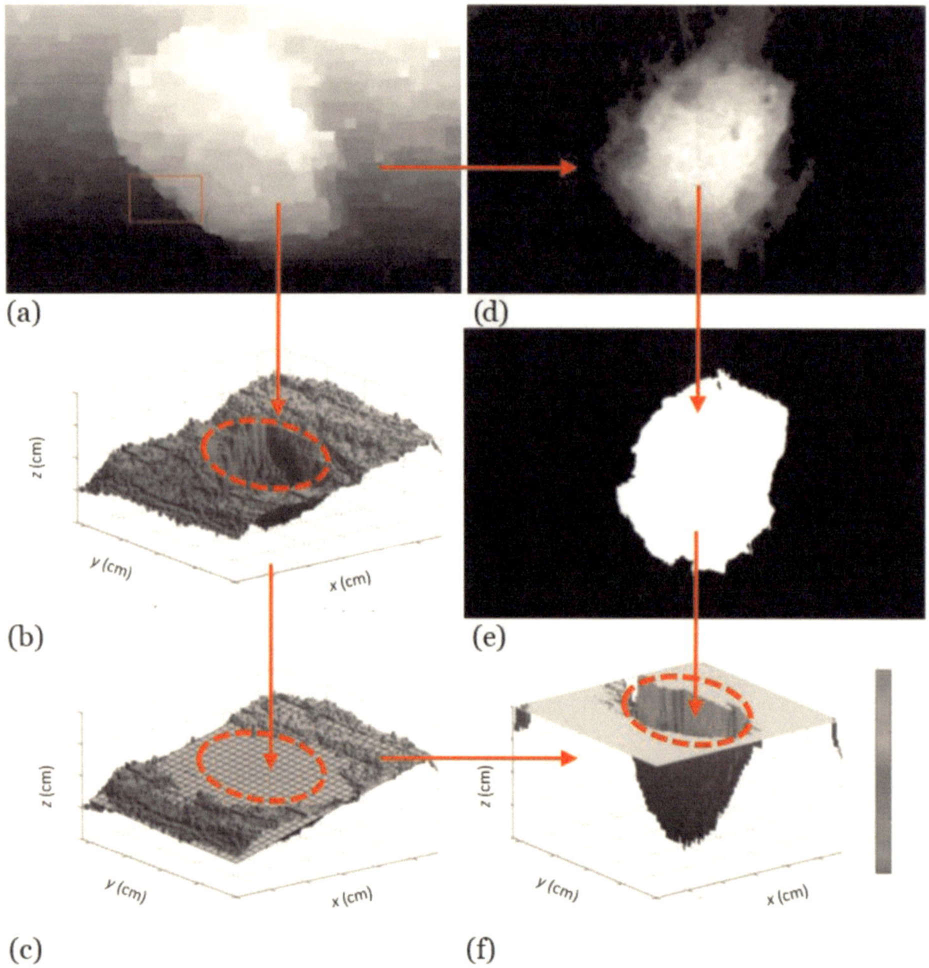

Figure 6.
(a) Pothole depth image. (b) Corresponding depth data to RGB image in (a). (c) Plane fitting using RANSAC [52]. (c) Relative depth obtained from subtracting the depth values from the fitted plane. (d) Rotated gray-scale representation of the relative depth values. (e) Detected pothole defect obtained from binarizing image (d) using the Otsu's thresholding. (f) Depth map of the detected pothole with dimensions in millimeters (cm).

edges by the color image (**Figure 9**). This implies that it is possible to improve the pothole detection from RGB imagery through fusion of the RGB and depth image datasets (feature fusion) or through post-segmentation fusion (object fusion). For this chapter, only a discussion and potential illustration is presented.

5.4 Evaluation of results and quantification of pothole metrology parameters

An evaluation of the low-cost pavement pothole detection system is carried out using 55 depth image frames comprising of 35 images with potholes and 20 defect-free frames were evaluated. The results of the illustrative evaluation are presented in **Tables 5** and **6**, respectively in terms of the confusion matrix and the overall performance indices: TP, TN, FP, and FN which respectively represent the true positive, true negative, false positive and false negative. In **Table 6**, accuracy is defined as the proportion of the true classifications in the test dataset, while precision is the proportion of true positive classifications against all positive

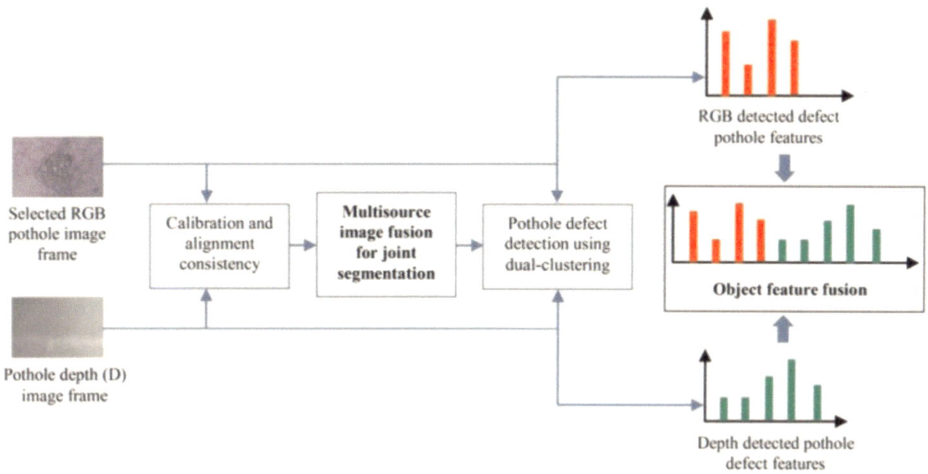

Figure 7.
Conceptual framework for the RGB-D pothole defect detection based on pre-detection image feature fusion and post-detection object fusion.

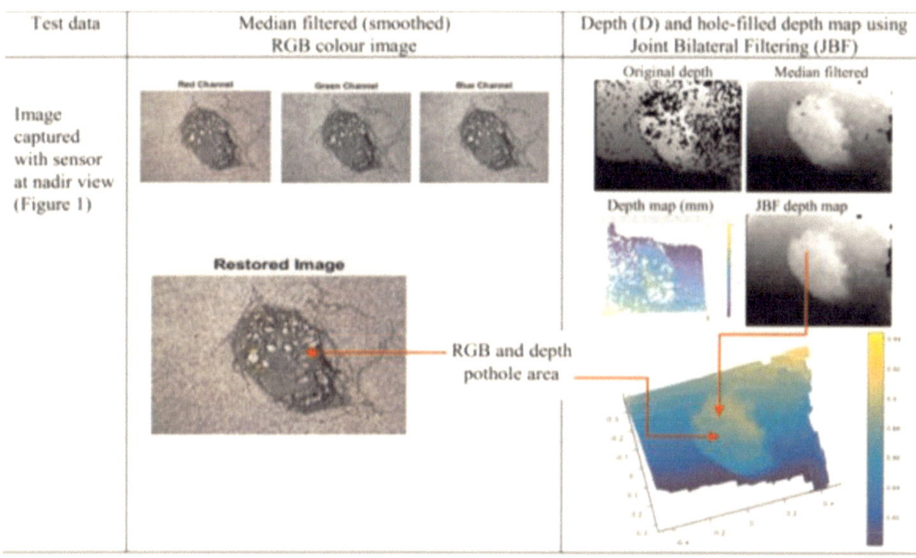

Figure 8.
Comparing RGB imagery (a) and filtered depth map for pothole and non-pothole mapping on asphalt pavement.

classifications. The overall results show that the detection rate for potholes was at 82.8% degree of accuracy.

In terms of the pothole metrology measurements, **Table 7** presents a sample summary of the results for the metrologic data quantification as characterized by: length and width, mean depth, mean surface area and volume of the potholes within image frames, and the resulting relative errors. From the results in **Table 7**, it is observed that while for some pothole defects the estimated dimensions are close to the ground-truth manual measurements, in few cases i.e., less 25% of the images, the relative error is more than 20%. This observed error magnitude in the pothole-detection system was attributed to the shape and edge complexity of the potholes, which are mathematically complex to represent and estimate appropriately and accurately as demonstrated in **Figure 6**.

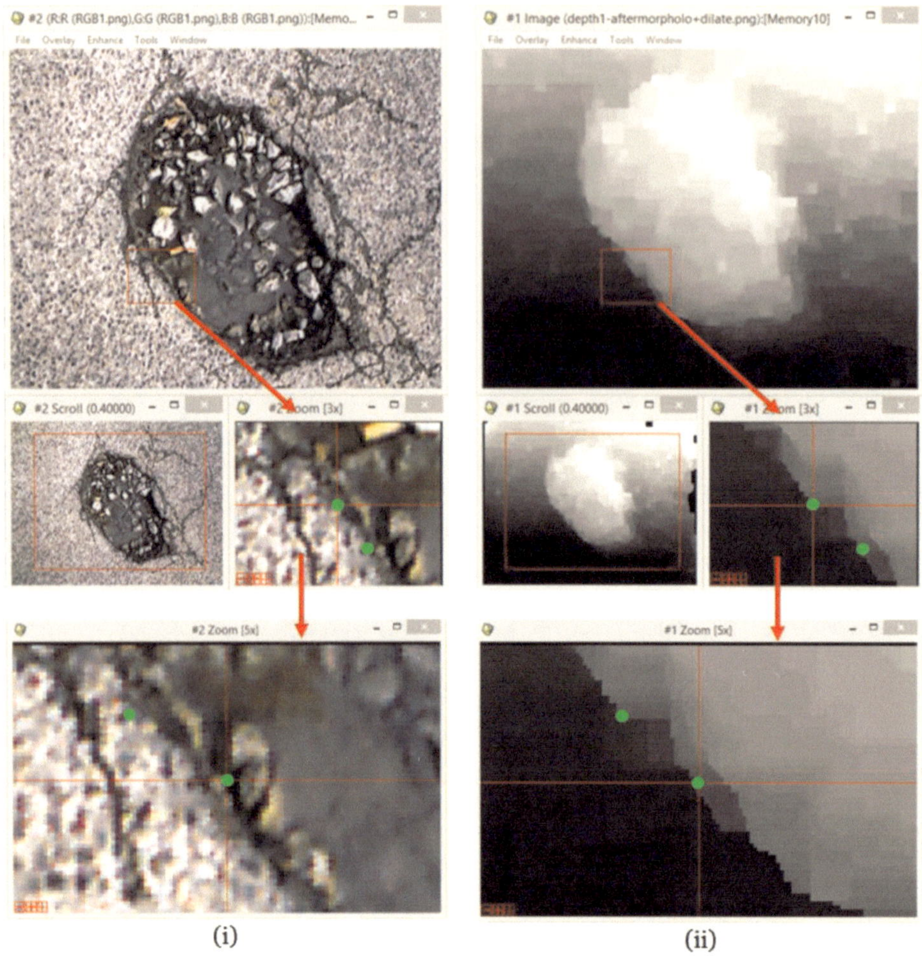

Figure 9.
Illustration of the significance of depth in pothole edge mapping in relation to pothole data fusion and improved detection. (i) RGB image. (ii) Depth map.

Prediction	Ground truth	
Classified	**Defective**	**Defect-free**
Defective (potholes)	TP = 40	FP = 5
Defect-free (non-potholes)	FN = 15	TN = 50

Table 5.
Confusion matrix of the evaluated pothole-defect detection system.

Index	Accuracy (%)	Precision (%)	Recall (%)
Value	82.8	88.8	72.7

Table 6.
Overall performance of the pothole-defect detection system.

Defect ID#	Ground-truth		Proposed method		Relative error		Proposed iMMSS method		
	Length (cm)	Width (cm)	Length (cm)	Width (cm)	Length (%)	Width (%)	Mean depth (cm)	Mean area (cm^2)	Volume (cm^3)
1	53.5	48.8	52.2	45.4	2.43	7.00	4.4	21.38	94.072
6	26.1	17.8	29.1	13.9	11.49	28.26	5.6	27.21	152.376
11	64.4	60.1	60.9	63.4	5.43	5.49	3.8	18.46	70.148
27	45.9	47.7	42.0	46.3	8.50	2.94	.59	28.66	169.094

Table 7.
Sample comparison of detected pothole metrological parameters with ground-truth measurements.

6. Conclusions

This chapter presents a robust approach for cost-effect detection of potholes on asphalt pavements. By first proposing a system for pavement surface mapping using Kinect v2.0 and based on the iMMSS hardware-software system, the implementation first incorporates k-means clustering and horizontal-vertical integration as data search or filtering algorithms, followed with spatial fuzzy c-means (SPCM) segmentation for pothole and non-pothole detection. The results of the processing illustrates the potential of using RGB and depth image in the detection of potholes based on low-cost consumer grade sensors, and shows the potential of fusing RGB + depth data for improved pothole detection.

From the experimental analysis, it is conclusive that using a single Kinect may not only limit the maximum traveling speed for data collection, but does not also cover the whole width of a traffic lane. This means that the field of view (FOV) can be increased by determining and using an array of Kinect sensors so that the lateral data collection extent can be increased. Further, the development of suitable depth and RGB fusion should be investigated both at object and at feature fusion levels.

In summary, it is demonstrated that low-cost and high-performance vision and depth sensors are capable of providing new possibilities for achieving autonomous inspection of pavement structures, and are suitable for overcoming the spatial and temporal limitations associated with both the manual human-based inspection and the expensive techniques. Overall, the findings of the study are significant, in terms of the new data and their processing challenges and results.

Acknowledgements

This research work was carried with the framework of research sponsorship by the Alexander von Humboldt Foundation (Germany), and the author would like to acknowledge and thank the Alexander von Humboldt Foundation for the financial support.

Author details

Yashon Ombado Ouma
Department of Civil Engineering, Geomatics Section, University of Botswana, Gaborone, Botswana

*Address all correspondence to: yashon.ouma@gmail.com

References

[1] Ouma YO, Hahn M. Wavelet-morphology based detection of incipient linear cracks in asphalt pavements from RGB camera imagery and classification using circular radon transform. Advanced Engineering Informatics. 2016;**30**(3):481-499

[2] Yan WY, Yuan X-X. A low-cost video-based pavement distress screening system for low-volume roads. Journal of Intelligent Transportation Systems: Technology, Planning, and Operations. 2018;**22**(5):376–389

[3] Huang J, Liu W, Sun X. A pavement crack detection method combining 2D with 3D information based on Dempster-Shafer theory. Computer-Aided Civil and Infrastructure Engineering. 2014;**29**(4):299-313

[4] Schnebele E, Tanyu BF, Cervone G, Waters N. Review of remote sensing methodologies for pavement management and assessment. European Transportation Research Review. 2015;**7**(2):1-19

[5] Zakeri H, Nejad FM, Fahimifar A, Torshizi AD, Zarandi MHF. A multi-stage expert system for classification of pavement cracking. In: IFSA World Congress and NAFIPS annual meeting (IFSA/NAFIPS), 2013 Joint. 2013

[6] Adu-Gyamfi Y, Okine NA, Garateguy G, Carrillo R, Arce GR. Multiresolution information mining for pavement crack image analysis. Journal of Computing in Civil Engineering. 2011;**26**(6):741-749

[7] Jahanshahi MR, Kelly JS, Masri SF, Sukhatme GS. A survey and evaluation of promising approaches for automatic image-based defect detection of bridge structures. Structure and Infrastructure Engineering. 2009;**5**(6):455-486

[8] Ouma YO, Hahn M. Pothole detection on asphalt pavements from 2D-colour pothole images using fuzzy *c*-means clustering and morphological reconstruction. Automation in Construction. 2017;**83**:196-211

[9] Zakeri H, Nejad FM, Fahimifar A. Image based techniques for crack detection, classification and quantification in asphalt pavement: A review. Archives of Computational Methods in Engineering. 2017;**24**(4):935-977

[10] Zhou J, Huang PS, Chiang F-P. Wavelet-based pavement distress detection and evaluation. Optical Engineering. 2006;**45**(2):027007

[11] Cord A, Chambon S. Automatic road defect detection by textural pattern recognition based on AdaBoost. Computer-Aided Civil and Infrastructure Engineering. 2012;**27**(4):244-259

[12] Werro P, Robinson I, Benbow E, Wright A. SCANNER accredited surveys on local roads in England – accreditation, QA and audit testing – annual report 2009–10. Wokingham, Berkshire: Transportation Research Laboratory; 2010

[13] Siddiqui AA. A new inspection method based on RGB-D profiling [MSc thesis]. Blacksburg, Virginia: Virginia Polytechnic Institute and State University; 2015

[14] Pöhlmann STL, Harkness EF, Taylor CJ, Astley SM. Evaluation of Kinect 3D sensor for healthcare imaging. Journal of Medical and Biological Engineering. 2016;**36**(6):857-870

[15] Pagliari D, Pinto L. Calibration of Kinect for Xbox one and comparison between the two generations of Microsoft sensors. Sensors. 2015;**15**:27569-27589

[16] Sell J, O'Connor P. The Xbox one system on a chip and kinect sensor. IEEE Micro. 2014;**2**:44-53

[17] Kolb A, Barth E, Koch R, Larsen R. Time-of-flight sensors in computer graphics. Computer Graphics Forum. 2010;**29**(1):141–159

[18] Mutto CD, Zanuttigh P, Cortelazzo GM. Time-of-flight Cameras and Microsoft Kinect. Springer Briefs in Electrical and Computer Engineering. New York: Springer-Verlag; 2012. Ch. 2. p. 21

[19] Shotton J, Sharp T, Kipman A, Fitzgibbon A, Finocchio M, Blake A, et al. Real-time human pose recognition in parts from single depth images. Communications of the ACM. 2013; **56**(1):116-124

[20] Khoshelham K, Elberink O. Accuracy and resolution of kinect depth data for indoor mapping applications. Sensors: Journal on the Science and Technology of Sensors and Biosensors. 2012;**12**(2):1437-1454

[21] Murthy SBS, Varaprasad G. Detection of potholes in autonomous vehicle. IET Intelligent Transport Systems. 2014;**8**(6):543-549

[22] Koch C, Jog G, Brilakis I. Automated pothole distress assessment using asphalt pavement video data. Journal of Computingin Civil Engineering. 2013; **27**(4):370-378

[23] Lokeshwor H, Das LK, Sud SK. Method for automated assessment of potholes, cracks and patches from road surface video clips. Procedia—Social and Behavioral Sciences. 2013;**104**: 312-321

[24] Parker R. Algorithms for Image Processing and Computer Vision. 2nd ed. Vol. 2011. New York, NY: John Wiley & Sons, Inc; 2011

[25] Wang P, Hu Y, Dai Y, Tian M. Asphalt pavement pothole detection and segmentation based on wavelet energy field. Mathematical Problems in Engineering. 2017;**2017**; Article ID 1604130, 13 pages. DOI: 10.1155/2017/1604130

[26] Zhang J, Hu J. Image segmentation based on 2D Otsu method with histogram analysis. Proceedings of Computer Science and Software Engineering, CSSE 2008. 2008;**6**: 105-108

[27] Xu D, Zhao P, Gui W, Yang C, Xie Y. Research on spectral clustering algorithms based on building different affinity matrix. In: 25th Chinese Control and Decision Conference (CCDC); 25–27 May. Vol. 2013. 2013. pp. 3160-3165

[28] Dunn JC. A fuzzy relative of the ISODATA process and its use in detecting compact well-separated clusters. Journal of Cybernetics. 1973; **3**(3):32-57

[29] Bezdek JC. Pattern Recognition with Fuzzy Objective Function Algorithms. MA, USA: Kluwer Academic Publishers Norwell; 1981

[30] Chuang K-S, Tzeng H-L, Chen S, Wu J, Chen T-J. Fuzzy C-means clustering with spatial information for image segmentation. Journal of Computerized Medical Imaging and Graphics. 2006;**30**:9-15

[31] Choudhry MS, Kapoor R. Performance analysis of fuzzy C-means clustering methods for MRI image segmentation. Procedia Computer Science. 2016;**89**:749-758

[32] Chen L, Lin H, Li S. Depth image enhancement for Kinect using region growing and bilateral filter. In: IEEE 21st International Conference on Pattern Recognition (ICPR). Vol. 2012. 2012. pp. 3070-3073

[33] Matyunin S, Vatolin D, Berdnikov Y, Smirnov M. Temporal filtering for depth maps generated by kinect depth camera. In: 3DTV Conference: The True Vision-Capture, Transmission and Display of 3D Video. 2011. pp. 1-4

[34] Camplani M, Salgado L. Efficient Spatio-temporal hole filling strategy for kinect depth maps. Proceedings of SPIE. 2012;**82900E**:2012

[35] Jung SW. Enhancement of image and depth map using adaptive joint trilateral filter. IEEE Transactions on Circuits and Systems for Video Technology. 2013;**23**(2):258-269

[36] Liu S, Wang Y, Wang J, Wang H, Zhang J, Pan C. Kinect depth restoration via energy minimization with TV21regularization. In: Proc. IEEE International Conference on Image Processing. 2013. pp. 724-724

[37] Tomasi C, Manduchi R. Bilateral filtering for gray and color images. In: Proceedings of the 6th IEEE International Conference in Computer Vision. 1998. pp. 839-846

[38] Yang Q, Ahuja N, Yang R, Tan K-H, Davis J, Culbertson B, et al. Fusion of median and bilateral filtering for range image upsampling. IEEE Transactions on Image Processing. 2013;**22**(12):4841-4852

[39] Petschnigg G, Agrawala M, Hoppe H, Szeliski R, Cohen M, Toyama K. Digital photography with flash and no-flash image pairs. ACM Transactions on Graphics. 2004;**23**(3): 664-672

[40] Eisemann E, Durand F. Flash photography enhancement via intrinsic relighting. ACM Transactions on Graphics. 2004;**23**(3):673-678

[41] Huber P. Robust Statistics. New York, NY, USA: Wiley; 1981

[42] Khoshelham K. Automated localization of a laser scanner in indoor environments using planar objects. In: Proceedings of International Conference on Indoor Positioning and Indoor Navigation (IPIN); 15–17 September 2010; Zürich, Switzerland. 2010

[43] Su PC, Shen J, Xu W, Cheung SC, Luo Y. A fast and robust extrinsic calibration for RGB-D camera networks. Sensors. 2018;**18**:235

[44] Bouguet J-Y. Camera Calibration Toolbox for Matlab. 2015. Available from: http://www.vision.caltech.edu/ -bouguetj/calib_doc

[45] Herrera D, Kannala CJ, Heikkilä J. Joint depth and color camera calibration with distortion correction. IEEE Transactions on Pattern Analysis and Machine Intelligence. 2012;**34**(10):2012

[46] Quintanilla-Dominguez J, Ojeda-Magaña B, Cortina-Januchs MG, Ruelas R, Vega-Corona A, Andina D. Image segmentation by fuzzy and possibilistic clustering algorithms for the identification of microcalcifications. Sharif University of Technology Scientia Iranica. 2011;**18**:580-589

[47] Moazzam I, Kamal K, Mathavan S, Usman S, Rahman M. Metrology and visualization of potholes using the Microsoft Kinect sensor. In: Proceedings of the 16th International IEEE Annual Conference on Intelligent Transportation Systems. Vol. 2013. 2013. pp. 1284-1291

[48] Koch C, Brilakis I. Pothole detection in asphalt pavement images. Advanced Engineering Informatics. 2011;**25**(3): 507-515

[49] Goldstein LJ, Lay DC, Schneider DI, Asmar NH. Calculus and its Applications. London: Pearson Education International; 2007

[50] Lyons L. Mathematics for Science Students. UK: Cambridge University Press; 2000

[51] Jahanshahi MR, Jazizadeh F, Masri SF, Becerik-Gerber B. An unsupervised approach for autonomous pavement defect detection and quantification using an inexpensive depth sensor. Journal of Computing in Civil Engineering. 2013;**27**(6):743-754

[52] Qian X, Ye C. NCC-RANSAC: A fast plane extraction method for Noisy range data. IEEE Transactions on Cybernetics. 2014;**44**(12):2771-2783